T0227557

Metal Recovery from Industrial Waste

Metal Recovery from Industrial Waste

Clyde S. Brooks

Contribution by:
Philip L. Brooks
George Hansen
Laurel A. McCarthy

CRC Press
Taylor & Francis Group
Boca Raton   London   New York

CRC Press is an imprint of the
Taylor & Francis Group, an **informa** business

First published 1991 by CRC Press
Taylor & Francis Group
6000 Broken Sound Parkway NW, Suite 300
Boca Raton, FL 33487-2742

Reissued 2018 by CRC Press

Library of Congress Cataloging-in-Publication Data

Brooks, Clyde S.
    Metal recovery from industrial waste / Clyde S. Brooks.
        p.   cm.
    Includes bibliographical references and index.
    ISBN 0-87371-456-3
    1. Nonferrous metals—Metallurgy.   2. Metal wastes—
    Recycling. I. Title
TN758.B74   1991
669.042—dc20   91-16911

A Library of Congress record exists under LC control number: 91016911

ISBN 13: 978-1-315-89535-2 (hbk)
ISBN 13: 978-1-351-07445-2 (ebk)

Visit the Taylor & Francis Web site at http://www.taylorandfrancis.com and the CRC Press Web site at http://www.crcpress.com

Preface

Metal Recovery from Industrial Wastes is an examination of technology with the potential for improving the economics of recovery and recycling of metals from industrial effluents. The separation technologies reviewed are in part novel and in part adaptations of existing processes at a mature state of development. The primary intent is to make an evaluation of the technical feasibility for separation of diverse multimetal wastes with widely varying chemical composition and physical properties. The objective is to identify the recovery opportunities not previously recognized and indicate the extent to which these have been reduced to practice. The relationship of metal recovery to management of hazardous wastes, resource conservation, economics, and future prospects is examined along with a review of the state of the art for relevant separation technology.

The appropriate audience for this book is

1. industrial management, engineers, and scientists actively involved with the development of metal recovery technology and/or hazardous waste disposal alternatives;
2. industrial consultants;
3. government personnel concerned with recycling and hazardous waste management; and
4. legislative personnel needing knowledge of the availability of technically feasible separation processes.

This book is offered as a practical primer on the hydrometallurgy of nonferrous metal reclamation from industrial wastes, as well as a brief discussion of pyrometallurgy, biological, and other separation processes. The objectives are to acquaint nontechnical readers with the subject as well as providing a detailed survey of current research and industrial practice to technical readers concerned with the management of industrial wastes.

About the Author

Clyde S. Brooks has 45 years experience in industrial research in applied physical chemistry, notably preparation of adsorbents and catalysts for fossil fuel processing, the surface chemistry of composites and industrial pollution control.

Mr. Brooks has worked on Koppers Company fellowships on coal chemical processing at the Mellon Institute of Industrial Research in Pittsburgh, PA and with the Exploration and Production Division of Shell Development Co. in Houston, TX on chemical stimulation of oil recovery. He was a senior research scientist at United Technologies Research Center in East Hartford, CT for 20 years, conducting research on a variety of corporate and government projects and doing internal consulting on fuel processing adsorbents and catalysts, surface chemistry of advanced materials (composites and rapidly solidified powders), and finding acceptable solutions for pollution problems for metal finishing wastes.

For the past 10 years Mr. Brooks has been an independent consultant (Recycle Metals), concentrating on developing technically feasible, economical technology for separation of nonferrous metals from industrial wastes and consulting on preparation and evaluation of adsorbents and catalysts. He has also had an important involvement in development of a technical assistance program for the Connecticut Hazardous Waste Management Service, assisting in selection of industrial awardees for matching state grants in pollution reduction and as a governor's appointee to a task force to develop a Connecticut state policy for landfill disposal of hazardous waste.

Mr. Brooks has a B.S. in Chemistry from Duke University, with graduate study at Duke University and Carnegie Institute of Technology (now Carnegie Mellon University). He was educated for industrial research on Koppers Company fellowships at the Mellon Institute of Industrial Research (now a department of Carnegie Mellon University).

He has over 70 publications and patents.

About Chapter 9 Authors

Putnam, Hayes & Bartlett is an economic and management consulting firm with offices in Cambridge, Massachusetts; London, England; Los Angeles, California; New York, New York; San Francisco, California; and Washington, DC. Members of the firms's consulting staff hold graduate degrees in business, economics, law, public policy, and technical disciplines. The backgrounds of the specific PHB staff who have participated in the preparation of this chapter are outlined below:

Philip L. Brooks (Washington, DC office) was a Principal until May 1990. He recently joined the Pittsburgh, Pennsylvania office of Arthur Andersen & Co. as Director of the Litigation Consulting Practice. In his eight years in consulting, Philip Brooks has worked extensively in the areas of regulatory economics, especially in the gas and environmental fields, litigation support, finance, accounting, and decision analysis. Mr. Brooks has managed the development of large and small financial models for damage estimation and the creation of large accounting and document tracking databases. In the environmental area, he has assisted steering committees with the allocation of cleanup costs at Superfund sites and evaluated features of and alternatives to the current Superfund legislation for several industry groups. Mr. Brooks received a B.A. in Biology from The University of Connecticut, an M.S. in pharmacotherapeutics from Long

Island University, and an M.S. in industrial administration from Carnegie Mellon University.

George Hansen (Washington, DC office) is an experienced analyst with a background in the management of environmental services projects. He has worked extensively in the analysis and allocation of Superfund and RCRA-related clean-up costs and in the financial evaluation of pollution control compliance expenditures. Mr. Hansen received a Bachelor of Engineering Science and Mechanics (B.E.S.M.) from the Georgia Institute of Technology and an M.S. in Industrial Administration (with Distinction) from Carnegie Mellon University.

Laurel A. McCarthy (Washington, DC office) is a Research Assistant with extensive research and economic modeling experience. Her project work has included computer modeling and financial and market analysis, as well as industry research in a number of different fields. She received a B.S. in Economics from the Wharton School of the University of Pennsylvania.

Acknowledgments

Particular acknowledgment goes to Philip L. Brooks, George Hansen, and Laurel A. McCarthy for preparation of Chapter 9 on Metal Recovery Economics. Mr. Brooks is the Director of the Litigation Consulting practice of Arthur Andersen & Co.'s Pittsburgh, PA office. Mr. Hansen and Ms. McCarthy are affiliated with the consulting firm of Putnam, Hayes & Bartlett, Inc. in Washington, DC and specialize in economic analyses.

Acknowledgment is due also to Philip Saxe of Innovative Support Services, Inc. in Wethersfield, CT for his conscientious efforts in assembly of the manuscript.

Contents

List of Tables

List of Figures

1

Introduction

The factors which facilitate creation of a favorable environment for the recovery of valuable nonferrous metals from industrial waste effluents are examined here. Principal incentives for recovery are reduction in the volume and toxicity of the waste effluents; the recovery of valuable metals; and the reduction of disposal costs.

In this chapter we consider sources of metal wastes, disposal alternatives and the current status of recycling. In Chapter 2 we examine the relevant regulatory framework. In Chapter 3 we examine the benefits that recycling provides to resource conservation. A state-of-the-art review of the technologies relevant to nonferrous waste metal separation and recovery is presented in Chapters 4 through 7. Table 1.1 summarizes the technologies reviewed. All of the nonferrous metals are considered, but principal emphasis is given to copper and nickel inasmuch as the amounts of these metals in commercial use are large enough to provide the most promise for favorable economics for recovery and recycling.

All of the technologies shown in Table 1.1 are hydrometallurgical in character except pyrometallurgy and biological processes. In Chapter 8 the technology discussion encompasses the demands of multimetal waste streams. Chapter 9 addresses the rapidly changing economics of metal recovery. Finally, in Chapter 10 an assessment is made of the prospects for recovery and recycling of nonferrous metals and identi-

Table 1.1   **Separation Alternatives**

Soluble metals
        Adsorption
        Cementation
        Electrowinning
        Ion Exchange
        Membrane Separations
        Precipitation
        Solvent Extraction

Solid wastes
        Biological Separations
        Flotation
        Magnetic Separations
        Pyrometallurgy
        Solvent Partition

fies some of the more promising avenues for pursuing recovery vs treatment and disposal.

1.1 SOURCES OF METAL WASTES

The sources of metal wastes are diverse in nature and in geographical distribution, as well as being impressively numerous. The volumes of all kinds of wastes generated on a nationwide scale are formidable. A national survey published in 1984, the Westat Report,[1] gave estimates of hazardous waste generators and treatment, storage, and disposal facilities regulated under the U.S. Resource Conservation and Recovery Act (RCRA) for 1981. More than 71 billion gallons, or 264 million tons, of hazardous waste of all sorts were generated by approximately 14,000 generators. More recent data communicated by Mike Burns of the Washington EPA office based on the EPA/RTI* contract in progress for the 1987 National Report of Hazardous Waste Generation and Treatment, Storage and Disposal Facilities Regulated Under RCRA, U.S. Environmental Protection Agency (1986

---

*RTI = Research Triangle Institute

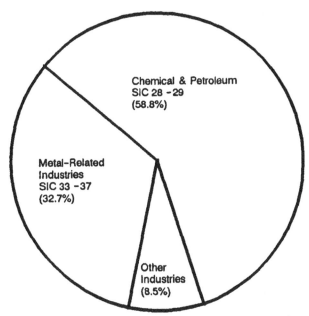

Figure 1.1. Percentages of hazardous waste generated by industry type. Based on 1987 National Report of Hazardous Waste Generation and Treatment, Storage and Disposal Facilities Regulated Under RCRA-EPA/RTI Contract (1986 data).

data) show some changes in these hazardous waste volumes and distributions by industry. The total waste reported now comes to 747 million tons because of changes in the way wastes are defined. There have also been changes in the relative distribution made by industry category (Figure 1.1). Chemical and petroleum industries (SIC 28–29) generate 58.8%, and metal-related industries (SIC 33–37) generate 32.7%, with 8.5% the remaining amount by other sources.

A survey of the hazardous waste generators in a small industrial state, Connecticut, was reported in 1989.[2] The report identified about 130,000 tons of all types of hazardous waste reported to the state's Department of Environmental Protection (DEP). The largest volume waste type, 64,700 tons or 43%, was inorganic solids/sludges containing metals generated by metal finishing, electronic, electrical, electroplating, machinery, and some chemical industries. These wastes were generated by the primary metals (SIC 33), fabricated metals (SIC 34), electrical and electronic machinery (SIC 36), and transportation equipment (SIC 37) industries.

1.2 DISPOSAL ALTERNATIVES

A hierarchy of waste disposal priorities commonly accepted consists of the following:

- Waste volume reduction
- Waste recycling
- Detoxification treatment
  - Physical
  - Chemical
  - Biological
- Incineration
- Solidification – stabilization
- Landfill
- Deep well injection

Metal recovery from waste effluents contributes to instrumentation of the first three of these categories.

The Westat study mentioned previously provides an overall view of current nationwide waste disposal practices of the 14,000 generators. The generators manage 96% of the waste on-site vs 4% off-site, with about 7% attributed to metal-related industries (SIC Codes 33–37). About 96% of hazardous waste is not recycled. Of the remaining 4% recycled, 81% is recycled on-site vs 19% off-site. Of the estimated total volume of hazardous waste generated in 1981 (71.3 billion gallons or 264 million metric tons), 48% was treated in some manner, 37% was stored and 15% was disposed of in surface impoundments, injection wells or landfills.

1.3 CURRENT STATUS OF RECYCLING

Nonferrous metals have the second priority for recycling,[3-5] after the precious metals. The nonferrous component of municipal solid waste is estimated to be about 2%, with aluminum the predominant metal. The amounts of the nonferrous metals, although minor components of municipal waste, are 2.0 million tons annually, with about 0.5 million tons of metals other than aluminum. The most promising potential for recycling of nonferrous metals undoubtedly lies with the metal finishing and electronics industries. The volumes of these wastes, while not providing new ore sources, offer the potential for yielding metal concentrations high enough to favor recovery processes. The wastes generated by the mining, mineral processing, and metallurgical industries in the United States are much more voluminous, and were estimated in 1987 to amount to 1.8 billion tons annually.[5,6] While not present at as high concentrations as many strategic metals from the metal-finishing industry, high-tonnage waste stockpiles exist in many locations in the western United States.

Commercial nonferrous metals that are significantly recycled as scrap are aluminum, copper and copper-base alloys, chromium, cobalt, cadmium, nickel, manganese, molybdenum, lead, titanium, zinc, and the precious metals.[7,8] The recycling of non-precious nonferrous metals seldom reaches 50%. Two million tons per year of aluminum, the principal nonferrous metal found in municipal waste, is currently recycled. This represents approximately 42% of annual consumption, up from only about 5 to 10% of the consumption in 1975.[4] Table 1.2 (from the U.S. Bureau of Mines Mineral Commodity Summaries for 1991) contains statistics on consumption, import reliance, and recycling for 22 metals of significant commercial use in the United States. The recycling of important nonferrous metals varies widely, ranging from negligible for arsenic, cadmium, magnesium, manganese, molybdenum, vanadium, and zirconium to 80% for titanium.[3,8,9] Recycling precious metals, except for silver,

Table 1.2   U.S. Metal Statistics, Consumption, Imports, Recycling

Metal	Consumption 1990 (10^3 tons)	Reliance on imports (% consumption)	Recycled metal Amount (10^3 tons)	Recycled metal % of consumption
Au	0.1	6.0	Negligible	–
Al	4800.0	98.0	2000.0	42.0
Sb	41.0	64.0	15.0	36.0
As	22.0	100.0	None	–
Cr	423.0	79.0	–	21.0
Co	8.0	86.0	1.1	13.0
Cd	3.0	54.0	Negligible	–
Cu	2200.0	5.0	530.0	24.0
Hg	*	*	0.2	13.0
Pb	1220.0	4.0	710.0	58.0
Mg	700.0	9.0	None	–
Mn	750.0	100.0	Negligible	–
Mo	17.0	Export	Negligible	–
Pt group (Rh, Ru, Ir, Os, etc.)	(98.8×10^3 kg)	88.0	(68.0×10^3 kg)	69.0
Ni	170.0	83.0	25.0	15.0
Ag	4.0	7.0	2.0	40.0
Sn	49.0	76.0	15.0	31.0
Ti (sponge)	25.0	Export	20.0	80.0
V	9.0	*	Negligible	–
W	9.0	73.0	2.0	21.0
Zn	1270.0	37.0	350.0	28.0
Zr	75.0	Export	Negligible	–

Source: U.S. Bureau of Mines – Mineral Commodity Summaries for 1991.
Notes: (1) Metals were selected on the basis that annual consumption was 1000 tons per year or more.
(2) * = withheld to avoid disclosing company propriety data.

ranges from 70 to 80% or better. Silver recycling comes to 40%. The sources used for citing the extent of recycling indicate that the information is not highly accurate.[3-10] The recycling of radioactive elements (thorium, uranium, etc.) is not public information and not considered here.

As can be seen in Table 1.2, even though the current percentages for several recycled metals are low, the total annual tonnages are not inconsequential; 530,000 tons of copper, 710,000 tons of lead, and 350,000 tons of zinc were recycled in 1990. For metals such as aluminum, antimony, copper, cobalt, cadmium, lead, magnesium, mercury, molybdenum, manganese, chromium, nickel, silver, vanadium, tungsten, zirconium, and zinc the potential for increasing the amount of recycling ranges from factors of two to five.[8-11]

It is useful when considering metal recycling to point out the magnitude of primary metal production and its persistent growth. Around the year 1900, annual metal production in the United States was about 18 million tons per year. By the year 1980, annual metal production increased to about 580 million tons per year.[10] The major metal wastes are quite large, amounting to several billion tons per year, principally generated in mining, milling, and primary production. Only a minor amount of this waste, 200 million to 300 million tons per year, appears as scrap for recycling.

Recycling provides several substantial advantages, one of which is the reduction in primary production waste for each ton recycled. For example, this amounts to 4 tons for iron, 200 tons for copper, and 200,000 tons for platinum each year. In addition, there may be savings in energy amounting to 50 to 90% of that required for primary production. Current scrap is 93% ferrous, with about 1% in stainless steel and the remaining 6% consisting principally of aluminum, copper, cobalt, chromium, lead, manganese, mercury, nickel, tin, titanium, tantalum, and zinc and the precious metals.[12] In spite of the intrinsic values of these nonferrous metals, obstacles to development of profitable recycling operations are formidable and include consumer use patterns, existing tax structures favoring raw materials over recycled materials, regulatory processes designed primarily for insulating society from toxic materials, and the lack of an infrastructure for handling recycled materials.

In the following chapters we will define some of the obstacles to recycling and the promising approaches for enhancing recycling. The principal emphasis is on a review of relevant separation technology which has promising potential for adaptation to technically feasible, economic recycling for the diverse, relatively small-scale waste streams of the electrical, electronic, machinery, and metal-finishing industries.

References

1. Westat, Inc. "National Survey of Hazardous Waste Generators and Treatment, Storage and Disposal Facilities Regulated under RCRA in 1981." EPA Office of Solid Waste Contract No. 68–01–6861 (1984).
2. "Connecticut Hazardous Waste Capacity Assurance Plan," Connecticut Hazardous Waste Management Service (October 1989), for submission to EPA by Connecticut DEP.
3. Spendlove, M. J. "Recycling Trends in the United States: A Review." U.S. Bureau of Mines Information Circular 8711 (1976).
4. "Base Line Forecasts of Resource Recovery, 1972 to 1990." U.S. Department of Commerce, NTIS PB 245924 (March 1975).
5. Hanna, H. S. and C. Rampacek. "Resources Potential of Mineral and Metallurgical Wastes," in *Fine Particles Processing*, Vol. 2, P. Somasundaran, Ed. (New York: AIME, 1980), p. 1709.
6. Hill, R. D. and J. L. Auerbach. "Solid Waste Disposal in the Mining Industry," in *Fine Particles Processing*, Vol. 2, P. Somasundaran, Ed. (New York: AIME, 1980), p. 1731.
7. *Study to Identify Opportunities for Increased Solid Waste Utilization*, Vol. 1 (New York: National Association of Secondary Material Industries, EPA-SW-40D. PB 212719, 1972), pp. 1–72.
8. "Technical Options for Conservation of Metals, Case Studies of Selected Metals and Products." Office of Technology Assessment, Congress of the United States,

Washington, D.C., OTA M-97, PB 80–102619 (September 1979).

9. Ness, H. "Industrial Wastes, Recycling Scrap – A Decade of Challenges and Frustrations," in *An Overview of the Sixth Minerals Waste Utilization Symposium*, S. A. Bortz and K. B. Higbie, Eds. U.S. Bureau of Mines Information Circular 8826 (1980), p. 82.

10. Spendlove, M. J. "Reclamation, Utilization, Disposal and Stabilization." U.S. Bureau of Mines Research on Resource Recovery, U.S. Bureau of Mines Circular 8750 (1977).

11. Herndon, R.C., Ed. *Proceedings of the First National Conference on Waste Exchange.* (Tallahassee, FL: The Florida State University, Florida Chamber of Commerce and U.S. EPA, March 1983).

12. Spoel, H. "The Current Status of Scrap Metal Recycling," *New J. Miner. Metals Mater. Soc.* 42(4):38 (1990).

2

Regulatory Considerations

The regulatory considerations that affect metal recycling consist principally of the pollution control measures for management of our water supplies and the hazardous waste handling regulations of the federal and state governments. In addition, there are tax laws governing the transport and management of raw materials, scrap, and recycled materials that can provide either barriers or incentives favoring recovery and recycling of waste metals.

Enlightened management of waste effluents has been the intent of a comprehensive body of federal and state legislation designed to adequately respond to public demand for an improved environment. Four federal laws constitute the principal regulatory framework, namely:

- Federal Water Pollution Control Act;
- Toxic Substance Control Act (TSCA);[1]
- Resource Conservation and Recovery Act of 1976 (RCRA);[2] and
- Comprehensive Environmental, Compensation and Liability Act of 1980 (CERCLA)[3]

The intent of TSCA, initially enacted in 1976, was to address the proper identification and accounting for toxic substances to protect public health. The intent of RCRA was to provide enlightened management for solid waste handling to both protect public health from toxic substances and conserve valuable natural resources. This law is the principal federal law providing incentives for recycling because the

11

ultimate goal is to minimize, if not eliminate landfill disposal. CERCLA, the fourth federal law and commonly known as the Superfund, has as its intent the cleanup and protection of the public health because of the nation's earlier ignorance and neglect in properly managing its waste effluents.

The regulatory considerations relevant to metal recycling will be reviewed briefly first. There are several recent publications[4-7] which deal comprehensively with the legal aspects of hazardous waste management. Initial guidance can be sought by contacting the appropriate state environmental agency (Appendix A), the appropriate EPA regional office (Appendix A), and the RCRA/Superfund hotline (Appendix A).

There are three categories of hazardous waste generators: large quantity, small quantity, and conditionally exempt. Large generators are responsible for an excess of 1000 kg of hazardous waste per month. A small-quantity generator produces

- more than 100 kg, but less than 1000 kg of nonacutely hazardous waste per month
- less than 100 kg per month of waste from cleanup of any residue or contaminated soil, water, or other debris from cleanup of acutely hazardous waste
- less than 1 kg per month of an acutely hazardous waste

Generators with smaller production rates than these are classified as conditionally exempt.

Solid wastes that are classified as hazardous must first meet the statutory RCRA definition as

> . . . any garbage, refuse, sludge from a waste treatment plant, water supply treatment plant, or air pollution control facility and other discarded material, including solid, liquid, semi-solid, or contained gaseous material resulting from industrial, commercial, mining, and agricultural operations, and from community activities, but does not include solid or dissolved material in domestic sewage, or solid or dissolved materials

in irrigation return flows or industrial discharges which are point sources subject to permits under section 402 or the Federal Water Pollution Control Act as amended (86 Stat. 880), or source, special nuclear, or byproduct material as defined by the Atomic Energy Act of 1954, as amended (68 Stat. 923.1). [RCRA, Sec. 1004, para. 27]

RCRA specifically excludes from definition of solid waste certain wastes generated by industrial processes [see CFR 261.4 (a)(1–7)]. Categories of waste of this type likely to contain waste metals would be industrial wastewater discharges that are point sources subject to regulation under section 402 of the Clean Water Act, as amended; special nuclear or byproduct material defined by the Atomic Energy Act of 1954, as amended, 42 U.S.C. 2011 et. seq.; and materials subject to in-situ mining which are not removed from the ground by an extraction process.

Additional metal wastes that are solid in nature, but excluded from being classified as hazardous wastes [see 40 CFR 261.4 (b), (c), and (d)] consist of:

- drilling fluids, produced waters, and wastes associated with exploration, development, or production of crude oil, natural gas, or geothermal energy;
- mining overburden returned to mine site;
- a number of specifically cited industrial wastes containing chromium, where the chromium is primarily trivalent and fails the characteristic EP toxicity test for chromium;
- solid wastes from extraction, benefication, and processing of ores and minerals (excluding coal), including phosphate rock and overburden from mining uranium ore; and
- analytical samples in transit or storage.

According to RCRA classification 40 CFR 261.2 (e) (i–iii), certain categories of recycled materials are not considered solid wastes if they are used or reused ingredients in an industrial process to make a product, provided the materials are not reclaimed, used, or reused as effective substitutes for commercial products or returned to the original process from which they were generated, without reclamation, or if the

Table 2.1 §261.31 Hazardous Metallic Wastes from Non-Specific Sources

Industry and EPA Hazardous Waste No.	Hazardous waste
F006	Wastewater treatment sludges from electroplating operations except from the following processes: (1) Sulfuric acid anodizing of aluminum; (2) tin plating on carbon steel; (3) zinc plating (segregated basis) on carbon steel; (4) aluminum or zinc-aluminum plating on carbon steel; (5) cleaning/stripping associates with tin, zinc, and aluminum plating on carbon steel; and (6) chemical etching and milling of aluminum.
F019	Wastewater treatment sludges from the chemical conversion coating of aluminum.
F007	Spent cyanide plating bath solutions from electroplating operations.
F008	Plating bath residues from the bottom of plating baths from electroplating operations where cyanides are used in the process.
F009	Spent stripping and cleaning bath solutions from electroplating operations where cyanides are used in the process.
F010	Quenching bath residues from oil baths from metal heat-treating operations where cyanides are used in the process.
F011	Spent cyanide solutions from salt bath pot cleaning from metal heat-treating operations.
F012	Quenching wastewater treatment sludges from metal heat-treating operations where cyanides are used in the process.
F028	Residues resulting from the incineration of thermal treatment of soil contaminated with EPA Hazardous Waste Nos. F020, F021, F022, F023, F026, and F027.

Table 2.2   §261.32 Hazardous Metallic Wastes from Specific Sources[a]

Industry and EPA Hazardous Waste No.	Hazardous waste
Inorganic pigments	
K002	Wastewater treatment sludge from the production of chrome yellow and orange pigments.
K003	Wastewater treatment sludge from the production of molybdate orange pigments.
K004	Wastewater treatment sludge from the production of zinc yellow pigments.
K005	Wastewater treatment sludge from the production of chrome green pigments.
K006	Wastewater treatment sludge from the production of chrome oxide green pigments (anhydrous and hydrated).
K007	Wastewater treatment sludge from the production of iron blue pigments.
K008	Oven residue from the production of chrome oxide green pigments.
Organic chemicals	
K021	Aqueous spent antimony catalyst waste from fluoromethanes production.
K022	Distillation bottom tars from the production of phenol/acetone from cumene.
K028	Spent catalyst iron from the hydroclorinator reactor in the production of 1,1,1-trichloroethane.
Explosives	
K044	Wastewater treatment sludges from manufacturing and processing explosives.
K045	Spent carbon from the treatment of wastewater containing explosives.
K046	Wastewater treatment sludges from the manufacturing, formation, and loading of lead-based initiating compounds.

Table 2.2 continued

Industry and EPA Hazardous Waste No.	Hazardous waste
Iron and steel	
K061	Emission control dust/sludge from the primary production of steel in electric furnaces.
K062	Spent pickle liquor generated by steel-finishing operations of plants that produce iron or steel.
Secondary lead	
K069	Emission control dust/sludge from secondary lead smelting.
K100	Waste leaching solution from acid leaching of emission control dust/sludge from secondary lead smelting.

[a]Modified by author to include only metallic wastes.

material is returned as a substitute for raw material feed stock where the process normally uses raw materials as principal feed stocks.

Metal containing solid wastes classified as hazardous unless specifically excluded under 40 CFR 260.20 and 260.22 are listed in Table 2.1, entitled Hazardous Metallic Wastes from Non-Specific Sources [see 40 CFR 261.31] and in Table

Table 2.3 Maximum Concentration of Metallic Contaminants for Characteristic of EP Toxicity

EPA Hazardous Waste No.	Contaminant	Maximum concentration (mg/L)
D004	Arsenic	5.0
D005	Barium	100.0
D006	Cadmium	1.0
D007	Chromium	5.0
D008	Lead	5.0
D009	Mercury	0.2
D010	Selenium	1.0
D011	Silver	5.0

2.2, Hazardous Metallic Wastes from Specific Sources [see 40 CFR 261.32].

The characteristics which determine classification as hazardous by the EPA consist of (1) ignitability, (2) corrosivity, (3) reactivity, and (4) EP toxicity. The fourth criterion, EP toxicity, is the primary concern for metal wastes. EP toxicity is a two-step process determined by laboratory analysis in which a sample of the waste is extracted and the extract is analyzed for contaminant concentration. The EPA hazardous waste categories for the metals and the maximum allowable concentration in the extract are given in Table 2.3, Maximum Concentration of Metallic Contaminants for Characteristic of EP Toxicity (40 CFR 261.24).

References

1. Toxic Substances Control Act, 15 U.S.C.A. SS2601–2629 (Publ: 94–580).
2. Resource Conservation and Recovery Act of 1976, 42 U.S.C.A. SS6901–6987 (Publ. L. 94–580).
3. Comprehensive Environmental Response Compensation, and Liability Act of 1980, 42 U.S.C.A. SS960–9657 (Publ. L. 96–510).
4. Lindgren, G.F. *Managing Industrial Hazardous Waste – A Practical Handbook* (Chelsea, MI: Lewis Publishers, 1989).
5. Bhatt, H.G., R.M. Sykes, and T.L Sweeney. *Management of Toxic and Hazardous Wastes* (Chelsea, MI: Lewis Publishers, 1989).
6. Phifer, R.W. and W.R. McTigue, Jr. *Waste Management for Small Quantity Generators* (Chelsea, MI: Lewis Publishers, 1988).
7. Bartlett, K.G. *Hazardous Waste Management*, (Chelsea, MI: Lewis Publishers, 1986).

3

Resource Conservation

Entirely aside from the merits of reducing environmental impact and minimizing the costs and hazards of waste metal disposal, there are the important considerations of the magnitude of our metal reserves and their strategic nature. Consideration of the United States' usage shows that there are about 20 metals with annual production in the United States in excess of 1,000 tons per year and 10 metals with production exceeding 50,000 tons per year.

It is necessary to examine how metal reserves are determined. Mineral reserves are defined by how much material can be mined at today's prices with today's technology. The two main economic factors which determine future mineral availability consist of:

1. geological extensions of supply through discoveries of new deposits and new zones in old deposits
2. technological extensions through increased ability to work lower-quality ores or substitution of unconventional materials

It is quite arbitrary to set an absolute limit on mineral resources in view of the wide distribution of minerals in the earth's crust. What determines a quantitative designation to a mineral reserve is the difficulty of finding the ore, the available recovery technology, and the cost in labor, and detrimental environmental effects accompanying recovery. Providing the costs of recovery in these various categories are assumed to be acceptable, reserves with progressively lower quality and metal content can be exploited.[1] This point of view is very much at variance with the Malthusian view of the Club of Rome's report on "The Limits of Growth," which predicts a collapse of society due to exhaustion of natural resources.[2]

Whether the pessimistic thesis of the Club of Rome[2] or the more hopeful points of view offered by other analyses[3] are accepted, recovery and recycling of metal resources from industrial wastes assume additional significance in providing savings in exploration costs, in the labor and energy costs of recovery, and in the associated environmental penalties. Recovery and recycling of waste metals acquire additional merit when consideration is given to the stress on technology and the increased environmental penalties of continually seeking recovery from more dilute concentrations from progressively more extensive areas.

A further factor in providing incentive for recovery and recycling of waste metals is their "strategic" nature. In the case of strategic metals, the national boundary becomes an additional geographic limitation on determination of available reserves. Some statistics on current United States consumption and import reliance are given in Table 3.1.[4] There are about 14 metals of industrial significance for which there is negligible production in the U.S. Among 37 metals and minerals in commercial use, the reserves in the U.S. represent less than 10% for about 22 metals and minerals, based on data complied by the U.S. Department of the Interior.[4]

Several efforts have been made to assess the magnitude of the rates at which metal resources of the United States are being exhausted. Two evaluations that are noteworthy are those of Lovering et al.[5] and Roper.[6] Roper, in estimating the times for attaining peak production in the United States, concluded that 20 metals had passed their peak production by 1975.

An interesting comparison is the relative mineral self-sufficiency of two nations with large resources, such as the U.S. and the U.S.S.R.[7] Both nations have large and varied resources, but the U.S.S.R. appears to have an advantage.

Table 3.1 1990 Net Import Reliance of Selected Nonfuel Mineral Materials as a Percent of Apparent Consumption[e,1,1]

Material	Percent	Major sources (1986–89)
Arsenic	100	France, Sweden, Chile, Mexico
Cesium (pollucite)	100	Canada
Columbium (niobium)	100	Brazil, Canada, Germany, Thailand
Graphite	100	Mexico, China, Brazil, Madagascar
Manganese	100	Rep. of South Africa, France, Gabon
Mica (sheet)	100	India, Belgium, France, Brazil
Rubidium	100	Canada
Strontium (celestite)	100	Mexico, Germany
Thallium	100	Belgium, U.K., Germany, France
Bauxite and alumina	98	Australia, Guinea, Jamaica, Surinam
Gem stones (natural and synthetic)	98	Belgium, Israel, India, Republic of South Africa
Diamond (industrial stones)	92	Ireland, U.K., Republic of South Africa
Asbestos	90	Canada, Republic of South Africa
Fluorspar	90	Mexico, Republic of South Africa, China, Spain
Platinum-group metals	88	Republic of South Africa, U.K., U.S.S.R.
Tantalum	86	Germany, Thailand, Brazil, Australia
Cobalt	85	Zaire, Zambia, Canada, Norway
Nickel	83	Canada, Norway, Australia, Dominican Republic
Chromium	79	Republic of South Africa, Turkey, Zimbabwe, Yugloslavia
Tin	76	Brazil, China, Indonesia, Malaysia
Tungsten	73	China, Bolivia, Germany, Peru
Stone (dimension)	70	Italy, Spain, Canada, Taiwan
Barite	69	China, India, Mexico, Morocco
Potash	68	Canada, Israel, U.S.S.R., Germany
Antimony	64	China, Republic of South Africa, Mexico, Hong Kong
Cadmium	54	Canada, Mexico, Australia, Germany
Selenium	54	Canada, U.K., Japan, Belgium-Luxembourg
Peat	47	Canada
Pumice and pumicite	40	Greece, Mexico, Turkey, Ecuador
Zinc	37	Canada, Mexico, Spain, Peru
Iodine	35	Japan, Chile
Gypsum	30	Canada, Mexico, Spain
Silicon	30	Brazil, Canada, Venezuela, Norway
Iron ore	26	Canada, Brazil, Venezuela, Liberia

Table 3.1 continued

Material	Percent	Major sources (1986–89)
Quartz crystal (industrial)	17	Brazil, Namibia
Nitrogen	14	Canada, U.S.S.R., Trinidad and Tobago, Mexico
Cement	13	Mexico, Canada, Spain, Greece
Sodium sulfate	13	Canada, Mexico
Iron and steel	12	European Community, Japan, Canada, Republic of Korea
Sulfur	11	Mexico, Canada
Salt	10	Canda, Mexico, Bahamas, Chile
Vermiculite	10	Republic of South Africa, China
Magnesium compounds	9	Greece, China, Canada, Ireland
Mica (scrap and flake)	8	Canada, India
Copper	5	Canada, Chile, Mexico, Peru, Zaire
Lead	4	Canada, Mexico, Australia, Peru

Source: Mineral Commodity Summary for 1991, U.S. Bureau of Mines.

eEstimated.

[1]Net import reliance = imports – exports + adjustments for Government and industry stock changes.

[2]Apparent comsumption = U.S. primary + secondary production + net import reliance.

The U.S.S.R is self-sufficient in 17 metals, namely, uranium, molybdenum, magnesium, titanium, iron ore, copper, lead, nickel, zinc, cobalt, mercury, chromium, manganese, tungsten, silver, gold, and platinum; whereas the U.S. is self-sufficient in 5 metals, namely, uranium, molybdenum, magnesium, titanium, and vanadium. A further consideration in regard to "strategic" materials is the national source of the imported materials in regard to political stability and orientation.[3-8] Many of the current sources of imports to the United States afford precarious political relationships.

There is a national policy in the U.S. for stockpiling strategic metals. And there is a potentially meaningful connection that could be made between metals recovered from wastes and the stockpiling of strategic metals if tax incentives could be provided to direct recovered materials into the "strategic" stockpile.

In spite of the arbitrary nature of the assumptions, it is useful to extrapolate current knowledge of demand vs resources to predict metals that will become short in supply within the foreseeable future (by about 2100). According to estimates by Holden,[9] metals that would be of particular concern as strategic metals for the United States consist of cobalt, chromium, magnesium, titanium, and the platinum metals. On the other hand, the important metals of commerce, such as iron, aluminum, manganese, magnesium, lead, zinc, chromium, nickel, and platinum, are not in critical supply on a worldwide basis, although some extensions of current recovery technology may be required to improve the economics.

A recent assessment by the U.S. Office of Technology Assessment[10] (OTA) arrived at a hierarchy of technical options for conservation of eight metals: iron, copper, aluminum, magnesium, chromium, nickel, tungsten, and platinum. This hierarchy consists of (1) substitution; (2) product recycling; (3) elimination of nonessential metal in products; (4) extended product life; and (5) reduced use of material-intensive systems such as alloys. The option of primary concern in this assessment is recycling. It is relevant to observe the current extent of recycling for the above cited metals, which comes to about 100% for iron and steel scrap, but 24% for copper, 42% for aluminum, negligible for magnesium, 21% for chromium, 15% for nickel, 21% for tungsten, and 69% for the precious metals.

The cited OTA study was initiated by the Committee on Commerce, Science and Transportation of the U.S. Senate as a result of the 1973 oil embargo focusing attention on the vulnerability of the U.S. in its dependence upon imported oil and other minerals, the increasing demand for materials arising from the economic recovery in 1973 to 1974 from the 1969 to 1970 recession, and the growth of U.S. per capita consumption of minerals to four times the world average, which stimulated an interest in reducing consumption through conservation.

Among the 35 metals listed in Bureau of Mines compilations,[4] most of which are in current commercial use in the United States, the following 22 metals have been selected for special review: aluminum, antimony, arsenic, cadmium, chromium, cobalt, copper, gold, lead, manganese, magnesium, mercury, molybdenum, nickel, platinum, silver, tin, titanium, tungsten, vanadium, zinc, and zirconium. All these metals are in annual use in the United States to the extent of 1,000 tons or more currently, with the exception of the precious metals. The selection of these particular metals is based primarily on three principal criteria: (1) limited U.S. reserves; (2) extensive industrial use; and (3) strategic importance.

Currently there is a relatively high level of recovery of metallic scrap for several of the principal commercial metals, such as iron, copper, lead, and aluminum, and the precious metals. However, particular attention is given in this review to metal wastes in the oxide/hydroxide state, in view of the significant amounts of certain of these metals generated as waste, their strategic significance, and, considering their difficult physical condition, the obvious need to develop adequate technology to permit economic recovery.

References

1. Brooks, D. B. and P. W. Andrews. "Mineral Resources, Economic Growth and World Population," *Science* 135:13 (1974).
2. Meadows, D. M., D. C. Meadows, J. Randers, and W. W. Behreus, III. *Report of the Club of Rome's Project on the Predicament of Mankind* (New York: Universe Books, 1972).
3. Goeller, H. G. and A. Zucker. "Infinite Resources: The Ultimate Strategy," *Science* 223:456 (1984).

4. *"Mineral Facts and Problems." U.S. Bur. Mines Bull.* 671 (1980).
5. Lovering, T. S. et al. *Resources and Man* (San Francisco: W. H. Freeman and Company, 1969).
6. Roper, L. D. *Where Have All the Metals Gone?* (Blacksbury, VA: University Publications, 1976).
7. Ewell, R. "Raw Materials," *Chem. Eng. News* (August 24, 1970), p. 43.
8. Lepowski, W. "Politics and the World's Raw Materials," *Chem. Eng. News* (June 4, 1979), p. 14.
9. Holden, G. "Getting Serious About Strategic Minerals," *Science* 212:305 (1981).
10. U.S. Office of Technology Assessment. "Technical Options for Conservation of Metals. Case Studies of Selected Metals and Products," PB80-102619–OTA Report M-97 (September 1979).

4

The Alternatives for Metal Separation and Recovery

There are a large number of separation processes applicable to recovery of metals from industrial wastes (Table 4.1). Many of these processes have reached a mature state of development and there are companies offering these recovery services. There are a number of processes that have potential application to secondary metal recovery that can be incorporated by waste generators as a component of their existing manufacturing processes. This technology review is primarily directed at this last approach.

The four chapters that follow provide a broad survey of the potentially applicable separation processes for a wide spectrum of nonferrous metals. The principal emphasis is on hydrometallurgical processes, especially in view of the energy savings advantages of low-temperature processing, the physical character of the waste streams, and the potential for the most favorable economics. However, consideration is also given to pyrometallurgical and biological processes when relevant to the principal emphasis on hydrometallurgy. The primary concern is processing of metals in aqueous or nonaqueous solution or insoluble form as oxide/hydroxide/carbonate/sulfide sludges or dusts, and secondarily the processing of metallic scrap.

There are several published reviews of technologies which are applicable to metal finishing industry wastes,[1-6] electroplating industrial practice,[7,8] water treatment,[9-13] recy-

Table 4.1 Separation Alternatives

Soluble metals	
	Adsorption
	Cementation
	Electrowinning
	Ion exchange
	Membrane separations
	Precipitation
	Solvent extraction
Solid wastes	
	Biological separations
	Flotation
	Magnetic separations
	Pyrometallurgy
	Solvent partition

cling,[14-16] and metal refining processes[17] that provide considerable detail about many of these separation-concentration processes. The author published a review of many relevant separation processes in 1986.[18] There are several recent books that provide access to pertinent separation technology.[19-23] In addition, there are reviews of the hydrometallurgical literature in the *Journal of Metals* of the American Metal Society and the British Institute of Mining and Metallurgy that permit updating of recent developments.[24-26]

The separation processes applicable to metals in aqueous, acid, or alkaline solution such as adsorption, cementation, electrowinning, ion exchange, membrane processes, precipitation, and solvent extraction are reviewed in Chapter 5.

The separation processes applicable to solid wastes consist of biological separations, flotation, magnetic separations, pyrometallurgy and solvent partition. These technologies are reviewed in Chapter 6. Solubilization of solid waste is given consideration in Chapter 7. In this chapter we examine the creation of soluble systems amenable to treatment by the processes covered in Chapter 5. Chapter 8 considers separation processes applicable to nonferrous metal recovery from multi-metal wastes and spent catalysts.

In excess of 1,000 individual publications have been examined in preparation of this review. At the start of the review a systematic retrieval was conducted of separation processes applicable to recovery of nonferrous metals, primarily copper and nickel, from industrial wastes using the *Chemical Abstracts* database for the 10-year period 1980 to 1990. The concentration on copper and nickel was influenced by the large volumes of these metals in commercial use and the prospects for providing recovered volumes sufficiently large as to provide favorable economics. The *Chemical Abstracts* search recovered a total of 629 literature citations. A second bibliography for recycling of nonferrous metals based on an IMMAGE database and published by the British Institute of Mining and Metallurgy in 1990 should also prove useful.[27]

References

1. "A State of the Art Review of Metal Finishing Waste Treatment," Water Pollution Control Research Series 12010 EIE (Columbus, OH: Battelle Memorial Institute, 1968).
2. Zievers, J. F. and C. J. Novotny. "Curtailing Pollution from Metal Finishing," *Environ. Sci. Technol.* 7(3):209 (1973).
3. Coulter, K. R. "Pollution Control and the Plating Industry," *Plating.* 57:1197 (1970).
4. Innes, W. P. and W. H. Toller, Jr. "Preliminary Steps in Waste Treatment," *Plating* 57:1205 (1970).
5. Watson, M. R. *Pollution Control in Metal Finishing* (Park Ridge, NJ: Noyes Data Corp., 1973).
6. "Summary Report: Control and Treatment Technology for the Metal Finishing Industry, In-Plant Changes," EPA 625/8–82–008 (1982).
7. Graham, A. K, Ed. *Electroplating Engineering Handbook* (New York, NY: Van Nostrand Reinhold Co., 1971).
8. *The Canning Handbook on Electroplating* (Birmingham, U.K.: W. Canning Ltd. (1978).

9. Berkowitz, J. B. et al. "Unit Operations for Treatment of Hazardous Industrial Wastes," NTIS PB 275054 and PB 275287 (1978).

10. Pojasek, R. B., Ed. *Toxic and Hazardous Waste Disposal*, Vol. 4 (Ann Arbor, MI: Ann Arbor Science Publishers, Inc., 1980).

11. "Environmental Pollution Control Alternatives: Sludge Handling, Dewatering, and Disposal Alternatives of the Metal Finishing Industry," EPA 625/5-82-028 (1982).

12. Dean, J. G., et al. "Removing Heavy Metals from Waste Water," *Environ. Sci. Technol.* 6(6):518 (1972).

13. Coleman, R. T. et al. "Sources and Treatment of Wastewater in the Non-Ferrous Metals Industry," EPA-600/2-80-074; NTIS PB80-196118 (1980).

14. Cochran, A. A. et al. "Development and Application of the Waste-Plus-Waste Process for Recovering Metals from Electroplating and Other Wastes," U.S. Bureau of Mines RI 7877 (1974).

15. Crumpler, E. P., "Management of Metal Finishing Sludges," EPA 530/SW/561; PB 263946 (1977).

16. Bhatia, S. and R. Jump. "Metal Recovery Makes Good Sense!" *Environ. Sci. Technol.* 11(8):752 (1977).

17. Biswas, A. K. and W. C. Davenport. *Extractive Metallurgy of Copper* (New York: Pergamon Press, 1976).

18. Brooks, C. S., "Metal Recovery from Industrial Wastes," *J. Metals*, 38(7):50 (1986).

19. Noll, K. E., C. N. Haas, C. Schmidt, and P. Koduala. *Recovery, Recycle and Reuse of Industrial Wastes in Industrial Waste Management Series*, J. W. Patterson, Ed. (Chelsea, MI: Lewis Publishers, 1985).

20. Taylor, A. R., H. Y. Sohn, and N. Jarrett, Eds. *Recycle and Secondary Recovery of Metals* (Warrendale, PA: The Metallurgical Society (AIME) 1985).

21. Patterson, J. W. and R. Passino, Eds. *Metals Speciation, Separation and Recovery* (Chelsea, MI: Lewis Publishers, 1987).

22. Van Lunden, J. H. L., D. L. Stewart, Jr., and Y. Sahai, Eds. *2nd International Symposium of Recycling Metals and*

Engineered Materials (Warrendale, PA: The Metals Society (TMS), Minerals, Metals and Materials Society, 1990).

23. *Recycling of Metalliferous Materials*, Proceedings of Conference of the Institution of Mining and Metallurgy (Birmingham, England, April 23–29, 1990) (Brookfield, VT: Institute of Mining and Metallurgy North American Publication Center, 1990).

24. Reddy, R. G., "Metal, Mineral Waste Processing and Secondary Recovery," *J. Metals* 39(4):34 (1987).

25. Doyle, F. M. "Developments in Hydrometallurgy," *J. Metals* 40(4):32 (1988).

26. Doyle, F. M. "Aqueous Processing of Minerals and Materials," *J. Metals* 41(4):51 (1989).

27. *Recycling of Non-Ferrous Metals, A Reference List* (London: Institute of Mining and Metallurgy, 1990).

5

Recovery Technology for Metals in Solution

This chapter reviews the technology for separation and recovery of metals from aqueous solution employing adsorption, cementation, electrowinning, ion exchange, membrane separations, precipitation, and solvent extraction.

5.1 ADSORPTION

Twenty two literature citations were obtained in the *Chemical Abstracts* search for applications of adsorption separation for copper and nickel from industrial wastes. Metals presented included Ag, As, Al, Be, Cd, Cu, Co, Cr, Fe, Hg, Mn, Ni, Pd, Sb, Sn, Tl, Th, U, and Zn (Table 5.1).

Adsorbents used consisted of inorganic materials like ceramic fibrous materials,[A-6, A-16] titania,[A-13] hydrotalcite,[A-17] cement,[A-14] metal salts,[A-11, A-14, A-15] TaS_2, V_2O_5,[A-10] an Al alloy,[A-11] and limestone.[A-12] A wide variety of organic materials[A-1, A-2, A-4, A-7, A-19, A-20, A-21] have been used. Carbon[A-3, A-5, A-8, A-9] and xanthate fibers,[A-18] along with alumina, are the most prominent adsorbents achieving significant commercial practice.

Metal removal efficiencies reported are commonly quite high, but without adsorbent regeneration. The organic adsorbents usually provide high metal recovery efficiency only by oxidation destruction or pyrolysis of the adsorbent.

The systems employing chemical regeneration of the adsorbent consisted of the use of NH_3 with Ni on a coal-

Table 5.1 Metal Separation by Adsorption

Waste system	Metals	Adsorbent	Adsorbent regeneration	Metal separation efficiency %	Ref.
Oxide-silicate ore	Ni	Coal-pyrite	3M NH_3	88 Ni	A-1
Cu wastes	Cu	Diisobutyl methane/Teflon®	0.1 N HCl	~100 Cu	A-2
Metal finishing wastes	Ag, Cu, Cd, Ni	Carbon (Svensuka Celulosa AB)	Destruction by ignition 600°C	High as Ag, CuO, NiO	A-3
Plating	Ni, Zn	Chelating resin			A-4
Metal finishing waste	Ag, Cu, Ni, Cd,	Carbon		High after ignition	A-5
Ni electrolysis waste	Ni	Fibrous Lavsan filter			A-6
Metal waste solution	Ag	Metal/polystyrene	Pyrolysis	95-99 Ag	A-7
Metal waste solution	Ag, Cu, Hg, Pd	Activated carbon/red P			A-8
Metal waste solution	Ni	Filtrasorb 400 carbon		55-98 Ni	A-9
Metal waste solution	Ag, Cu	TaS_2, V_2O_5 intercalates		Low for low conc. Ag, Cu	A-10
Metal waste solution	Cu	Al alloy, chlorides present			A-11
Acid wastewater	Al, Cu, Fe, Zn	Na-di-Me-dithiocarboamate + limestone		High for Al, Cu, Fe	A-12
Radioactive waste	Cu, Co, Mn, Ni	Hydrated titania	No	Low after sintering	A-13

Heavy metal waste	Cu, Pd, Zn	Portland cement + chlorides		High in pelleted product	A-14
Cu wastewater	Cu	Cation exchange + chelating fiber			A-15
Spent electroless solution	Cu, Ni	Inorganic fiber (mica) and amino compound	No		A-16
Metal wastes	Ag, As, Be, Cd Cu, Cr, Hg, Ni, Sb, Sn, Pb, Ti, Zn	Hydrotalcite	0.01–1% alkali metal hydroxide		A-17
Industrial wastewater	Ag, Cd, Cu, Hg	Cellulose xanthate	Thermal or chem. oxid. (NaOCl)		A-18
Electronic waste	Ni, Pd	Cascein		Separates Ni and Pd	A-19
Ore wastewater	Co, Cu	Lignite			A-20
Radioactive waste	U, Th	Persimmon extract		High	A-21
Wastewater	Al, Cd, Co, Cr, Cu, Fe, Mn, Ni, Pb, Zn	MgO	H_2O/EDTA	94–98 all metals	A-22

pyrite,[A-1] HCl with Cu for a diisobutyl methanol Teflon adsorbent,[A-2] alkali metal hydroxides for various metals on active carbon plus hydrotalcite,[A-17] the use of chemical oxidation with NaOCl for Ag, Cu, Cd, and Hg on cellulose xanthate,[A-18] and aqueous ethylene diamine tetraacetic acid (EDTA) on MgO.[A-22]

Chemical and physical adsorption separation removal from dilute aqueous solutions on the low-cost, nonregenerable adsorbents may prove to be an attractive cleanup process for the nonprecious metals. If the adsorbents are in any measure regenerable it is probably because the attachment mechanism is an ion exchange process. Concentration from dilute solution on substrates which can be destroyed by combustion can be an attractive alternative for recovery of the precious metals from dilute solutions.

References

A-1.   Ferreira, R. C. H. "Recovery of Nickel Oxide Ore and Extraction of Nickel with Coal Containing Sulfur or with Wastes of Coal Beneficiation Plants or with Coal-Pyrite Mixtures." Brazilian Patent 77705715 (1979).

A-2.   Agency of Industrial Sciences and Technology. "Recovery of Copper." Japanese Patent 56019384 (1981).

A-3.   Echigo, T. "Removal and recovery of heavy metals by SCA carbon," *Kogai* 16(4):197–207 (1981).

A-4.   Matsuba, T. "Recovery of Heavy Metals from Wastewater with Chelating Resin," *Kagaku Sochi* 23(10):141–145, 98 (1981).

A-5.   Echigo, T. "Metal Recovery from Wastewater by Carbon Adsorbent," *Nenpo—Fukui-ken Kogyo Shikenjo*, Volume Date 1979:104–119 (1980).

A-6.   Nekhoroshkin, G. F., S. D. Barkov, and Y. D. Kolbas. "A Filter for Purifying Aspiration Gases from

Hydrometallurgical Production," *Tsvetn. Met.* (7):40–43 (1981).

A-7.   King, J. R., Jr. "Metal Removal Apparatus and Method." U.S. Patent 4331472 (1982).

A-8.   Lehr, K., G. Heymer, C. May, and H. Klein. "Winning Metals from Aqueous Solutions." European Patent 52253 (1982).

A-9.   Gianguzza, A. and S. Orecchio. "Use of Activated Carbon in the Treatment of Waters Containing Heavy Metals," *Inquinamento* 24(7–8):39–39 (1982).

A-10.   Celik, M. C. and D. J. Fray. "Intercalation as a Means of Metal Refining," *Trans. – Inst. Min. Metall.* Sect. C. 91:171–176 (1982).

A-11.   Miura, K. "Metal Recovery from Aqueous Solutions." Japanese Patent 59190333 (1984).

A-12.   Kobayashi, S., S. Okado, and Y. Yamada. "Treatment of Wastewater Containing Valuable Metals." Japanese Patent 60139387 (1985).

A-13.   Fuijiki, Y., T. Sasaki, and Y. Komatsu. "Immobilization of Divalent Transition Metals from Aqueous Solutions Using Crystalline Hydrated Titania Fibers," *Yogyo Kyokai Shi.* 93(5):225–229 (1985).

A-14.   Taguchi, Y. "Mixing Agent for Forming Adsorbent for Recovery Heavy Metal from Waste Ore." Japanese Patent 61133124 (1986).

A-15.   Goto, M. and S. Goto. "Removal and Recovery of Heavy Metals by Ion Exchange Fiber," *J. Chem. Eng. Jpn.* 20(5):467–472 (1987).

A-16.   Kanbe, T., Y. Kumagai, K. Urabe, H. Sugihara, and S. Sugiyama. "Utilization and Recovery of Useful Ingredients in Waste Electroless Coating Solutions," *Jitsumu Hyomen Gijutsu* 34(8):286–293 (1987).

A-17.   Sood, A. "Process for Removing Heavy Metal Ions from Solutions Using Adsorbents Containing Activated Hydrotalcite." U.S. Patent 4752397 (1988).

A-18.   Tiravanti, G., D. Marani, R. Passino, and M. Santori. "Synthesis and Characterization of Cellulose Xanthate Chelating Exchangers for Heavy Metal Removal and Recovery from Wastewaters," *Stud. Environ. Sci.*, 34 (*Chem. Prot. Environ.*, 1987): 109–118 (1988).

A-19.   Holpuch, V. and M. Milner. "Recovery of Palladium from Wastewater Printed Circuit Electroplating," *Metalloberflaeche* 43(5):224–226 (1989).

A-20.   Saleeva, N. V. and M. K. Kydynov. "New Natural Sorbent and the Potentials of Its Use," *Izv. Akad. Nauk Kirg. SSR, Khim-Tekhnol. Nauki* (1):85 (1989).

A-21.   Sakaguchi, T. and A. Nakajima. "Recovery of Uranium by Biological Substances" (Proceedings of the International Conference Separation Science and Technology, 2nd Canada: Ottawa, M. H. I. Baird and S. Vijayan, Eds.) *Can. Soc. Chem. Eng.* 1:331–336 (1989).

A-22.   Khalafalla, S. E., S. E. Pahlman, and D. N. Tallman. "Reclaiming Heavy Metals from Wastewater with Magnesium Oxide," in *Recycle and Secondary Recovery of Metals*, P. R. Taylor, H. Y. Sohn, and N. Jarrett, Eds. (Warrendale, PA: Metallurgical Society 1985).

5.2 CEMENTATION

Cementation is a form of precipitation in which there is an electrochemical mechanism (Tables 5.2 and 5.3). A metal with a more positive oxidation potential will pass into solution to replace a metal with a less positive potential[C-1] (Table 5.3). Copper is the metal most commonly separated by cementation, but the precious metals, Ag, Au, and Pd, as well as As, Cd, Ga, Pb, Sb, and Sn, have also been so separated. The metals of greater positive potential used, usually in the form of scrap or waste powders are aluminum, copper, iron, magnesium, or zinc. High separation efficiencies ranging from 70% to in excess of 99% are obtained for a wide variety of waste metal systems (Table 5.2).

Table 5.2 Metal Separation by Cementation

Waste system	Metals	Cementation metal	Metal separation efficiency %	Ref.
Metal waste	Cu	Al		C–5
Flue dust	Cu	Fe	91–98 Cu + 21 As	C–6
Wastewater	Cu	Fe	0.1 ppm Cu residual conc.	C–7
Acid wastewater	Cu	Fe or Al rings	70 Cu on Fe; 95 Cu on Al	C–8
Metal powder reclamation	Cu	Fe (rotating disk)	99.6 Cu (2 stage)	C–9
Cu waste	Cu	Fe		C–10
Waste sludge	Cu, Cr, Zn	Fe for Cu/Zn	98.9 Cr	C–11
Galvanic, phosphatic sludge	As, Cu, Sb, Sn	Zn		C–12
Etching waste	Cu	Zn	99 Cu and Zn	C–13
Alloy scrap	Ag, Cu, Pd	Cu for Ag/Pd	99.96 to 99.99 Ag/Pd	C–14
Flux waste	Cu, Pb	Zn	99.5 Cu	C–15
Electronic scrap	Ag, Pd	Cu		C–16
Acid ore heap leachate	Cu	Fe		C–17
Fe/Cu chloride solutions	Cu	Fe		C–18
Acid metal solution	Ag, Au, Cu, Pb, Ga	Fe for Cu; Fe or Zn for Pb; Zn for Ag/Cu; Cd for Au		C–1
Acid metal solution	Cu	Fe		C–2
Acid metal solution	Cu	Fe		C–3
Metal wastes	Cd	Mg	High Cd removal	C–4
Hydrometallurgy waste	Cu	Fe	86 Cu	C–19
Plating	Cu	Al		C–20
Cu wastewater	Cu	Al		C–21
Cu leachate	Cu	Fe		C–22

Table 5.3 Cementation Systems: Electrical Potential at 25°C for 1 N Solution

Substrate Metal	Volts	Metal deposited	Volts
Mg/Mg^{2+}	+2.37	Cd/Cd^{2+}	+0.40
Al/Al^{3+}	+1.66	Ni/Ni^{2+}	+0.25
Zn/Zn^{2+}	+0.76	Sn/Sn^{2+}	+0.14
Fe/Fe^{2+}	+0.44	Pb/Pb^{2+}	+0.13
Fe/Fe^{3+}	+0.36	$Cu/Cu^{2=}$	−0.34
		Ag/Ag^{2+}	−0.80
		Pd/Pd^{2+}	−0.99
		Pt/Pt^{2+}	−1.2
		Au/Au^{3+}	−1.5

References

C-1.   Habashi, F. *Principles of Extractive Metallurgy, Vol. 2, Hydrometallurgy* (New York: Gordon & Breach Science Publishers, Inc. 1970).

C-2.   Jester, T. D. and T. H. Taylor, in *Proceedings of the 28th Purdue Industrial Waste Conference* (LaFayette, IN: Purdue University Press, 1973), p. 129.

C-3.   Patterson, J. W. and W. A. Sancuk, in *Proceedings of the 32nd Purdue Industrial Waste Conference* (LaFayette, IN: Purdue University Press, 1977), p. 853.

C-4.   Gould, J. P., B. Khudenko, and H. F. Wiedeman. "The Kinetics and Yield of the Magnesium Cementation of Cadmium," in *Metals Speciation, Separation and Recovery*, J. W. Patterson and R. Passino, Eds. (Chelsea, MI: Lewis Publishing Inc., 1987), p. 175.

C-5.   Murr, L. E., V. Annamaiai, and P. C. Hsu. "A Hydro-Saline (Chloride-Ion) Cycle for Copper-Bearing Waste Leaching," *J. Metals* 31(2):26–32 (1979).

C-6.   Prater, J. D. and B. A. Wells. "Recovery of Copper from Arsenic-Containing Metallurgical Waste Materials." U.S. Patent 4149880 (1979).

C-7.   Kubo, K., A. Mishima, T. Aratani, and T. Yano. "Copper Recovery from Wastewater by Cementation

Utilizing Packed Bed of Iron Spheres," *J. Chem. Eng. Jpn.* 12(6):495–497 (1979).

C-8.   Ambartsumyan, A. K. and G. P. Gzraryan. "Device for Continuous Copper Recovery from Acid Copper-Containing Wastewater," *Khim. Promst. (Moscow)* (5):310–311 (1980).

C-9.   Episkoposyan, M. L., S. S. Bakhchisaraitseva, S. K. Karpetyan, O. N. Shakhbazyan, V. T. Aivazyan, and A. N. Karibyan. "Production of Metal Powders from Reclamation Solutions," *Promst. Arm.* (10):31–34 (1982).

C-10.   Sedahmed, G. H. and M. A. Fawzy. "Electrowinning of Copper Via Galvanic Cementation in a Cell Using Fixed Bed Electrodes and a Static Solution," *Bull. Electrochem.* 1(3):281–283 (1985).

C-11.   Ghetie, F. N., S. Chiriac, and I. Mitiu. "Metal Recovery from Sludges Resulting in Chemical Wastewater Treatments." Romanian Patent 88174 (1985).

C-12.   Kugler, S. and G. Szalai. "Nonselective Processing of Galvanic and Phosphatic Sludges." Hungarian Patent 41263 A2 (1987).

C-13.   Berkesi, J. and T. Gilanyi. "Process for Protecting the Environment from Acidic and/or Basic Etching Waste." Hungarian Patient 41447 (1987).

C-14.   Perte, E., M. Marc, O. Ceuca, G. Cosmina, O. Crucin, and L. Pacuraru. "Recovery of Copper, Palladium, and Silver from Alloy Scrap." Romanian Patent 90071 B1 (1986).

C-15.   Akerlow, E. V. "Process for Recovering Metals and Metallic Salts from Flux Wastes." U.S. Patent 4756889 (1988).

C-16.   Loebel, J. and L. Meissner. "Recovery of Noble and Accompanying Metals from Electronic Scraps." German Democratic Republic Patent DD 253048 (1988).

C-17.   Druzhinina, S. I., D. A. Pirmagomedov, V. L. Aranovich, B. B. Beisembaev, and B. K. Kenzhaliev. "Utilization of Acidic Copper-Containing Wastes in Heap Leaching of Low-Grade Copper-Porphyric

Ores." *Kompleksn. Ispol'z. Miner. Syr'ya.* (8):49–52 (1988).

C-18. Mikhailovskii, V. L., V. E. Ternovtsev, Y. S. Sergeev, L. A. Gergalov, S. V. Sokolov, and S. M. Novak. "Regeneration of Iron-Copper Chloride Solutions." U.S.S.R. Patent 1435660 (1988).

C-19. Tomasek, K., E. Kassayova, L. Molnar, S. Cempa, P. Vadsaz, E. Reitznerova, J. Simko, and L. Weigner. "Hydrometallurgic Method of Copper Extraction from Copper Works Waste." Czechoslovakian Patent 234897 (1987).

C-20. Pekarskii, L. D., E. V. Kazakov, S. A. Ushakov, and M. L. Petrov. "Manufacture of Copper Powder." U.S.S.R. Patent 1285037 (1987).

C-21. Hiromasa, M. "Recovery of Copper from Wastewater with Aqueous Aluminum Salt Production." Japanese Patent 61205617 (1986).

C-22. "Copper Powder By Recycling." Japanese Chemicals Inspection and Testing Institute. Japanese Patent 60125304 (1985).

5.3 ELECTROWINNING

Electrowinning is a mature technology and has been used extensively for separation and recovery of metals from solution. The 136 publications reported on for this separation process include a wide variety of waste systems ranging from the heap leachates of extractive hydrometallurgy, mine leachates, plating wastes, spent electroless solutions, a variety of hydrometallurgical waste solutions, solubilized dusts and sludges, a variety of waste acids, a variety of metal finishing wastes, electronic and electrical wastes, and metal scrap. (See Table 5.4.)

There are a number of general references on electrochemical deposition that are a good starting point for the fundamental information needed for adaptation to metal recovery from wastes.[E-1, E-2, E-18, E-8] There are also useful publications

Table 5.4 Metal Recovery by Electrowinning

Waste	Metals	Special features	Recovery efficiency %	Ref.
Cyanide liquor	Zn	Anode of conductive graphite particles	99 Zn	E-33
Wastewater	Ag, Au, Cd, Cu, Ni, Sn, Zn	Vibrating cell with S present	~99 of metals	E-39
Cu scrap	Cu,	High purity Cu power product	to 99 Cu	E-40
Metal wastes	Ag	3-Dimensional electrode	>99 Ag	E-42
Metal wastes	Au, Cu, Ni, Zn	Fluidized glass beads		E-52
Metal wastes	Cu	Graphite electrode, oxidative degradation of organics accompanies electrolysis		E-64
Plating wastes	Cd, Cu, Ni, Zn	Bi-polar disk electrodes		E-67
Metal wastes	Precious Metals and Zn	Particulate flow-by electrode		E-68
Filter dust and sludge	Co, Cd, Ni	Cd recovered by electrolysis		E-71
Metal dusts	Cd, Cu, Pb	$CdCO_3$, Pb + Zn sep by volatilization, leaching, cementation	99.95 Cu	E-72
Plating waste	Ni	An improved process with water recycling	94	E-75
Metal waste	Ag, Cu, Cr, Ni	Fluidized bed electrolysis using chelated agents to concentrate metals		E-77
Brass smelter dust	Pb, Zn	Spent battery components used to form electrolysis cell	80–95 Pb/Zn	E-78

Table 5.4 continued

Waste	Metals	Special features	Recovery efficiency %	Ref.
Cu wastes	Cu	Process uses electrodialysis cell and a filter with magnetic granules	98 Cu	E-81
Ag wastes	Ag	Continuous Ag recovery for dilute solutions with cyanide		E-85
Cu wastes	Cu	Use of ultrasonics and chelating agent–steel cathode/ferrite anode	99.86 Cu powder	E-87
Plating wastes	Ni	SUS stainless 304 anode/Pt-coated Ti cathode–aided by presence Na-citrate		E-89
Galvanizing waste	Cu	Small system ideal for use presence of $Na_2S_2O_3$, NaOH, NaCl, EDTA, CN^-	99.98 Cu	E-91
Cu scrap	Cu	Cu scrap without melting put in electrical conducting Ti or stainless basket	99.95–99.98 Cu	E-95
Plating wastes	Ni	Cell had ultrasonic magnostrictive transducer		E-102
Sn plate/Cu scrap	Sn, Cu	Leach and electrolysis	90.3–99.3 Sn	E-106
Cd/Ni waste	Cd, Ni	Fluidized bed electrolysis	>99 Cd/Ni	E-113
Ni wastes	Ni	Low conc. Ni recovery with rotating tubular bed reactor and ion exchange		E-115

from the extractive metallurgy literature,[E-7, E-10 to E-13, E-27] applicable to the metals of interest: cadmium, cobalt, copper, chromium, nickel, lead, tin, zinc, etc. There are now a number of publications on work specifically directed to removal, if not metal recovery, by electrowinning from aqueous metal waste solutions[E-3 to E-6, E-9, E-14, E-17, E-19 to E-26, E-28, E-122] that are appropriate for treatment of hazardous wastes. In addition, there are commercially available electrowinning systems provided by Chemlec,[E-36] Lancy,[E-43] etc. (see Chapter 10, Table 10.1).

There are a number of relevant improvements in apparatus and operating procedure directed to more efficient metal recovery from dilute metal solutions[E-9, E-15] by using fluidized bed electrodes,[E-16, E-46, E-68, E-76, E-112] various moving electrode configurations,[E-32, E-38, E-50, E-52, E-57, E-61, E-66] modifications of the composition of the electrodes,[E-30, E-33 to E-35, E-45, E-46, E-49, E-61, E-64, E-79, E-88, E-98, E-102] extended area electrodes,[E-42, E-59] adjuncts such as ultrasonic vibrations,[E-86, E-101] and bubbled gas or agitated particles external to the electrodes.[E-31, E-38, E-52, E-67, E-80, E-119] Metals are most commonly deposited as removable sheets, but some systems provide high-purity powders.[E-40]

Waste systems that provide metal concentrations of significance, approaching 1 wt%, can be handled with conventional commercial electrodeposition apparatus. The more dilute systems require modifications such as indicated above to achieve high-efficiency metal separation and low residual metal concentrations in the effluent without excessive power consumption and severe diffusion limitations.

Electrochemical techniques employing membranes such as electrodialysis for metal recovery are covered in the section on Membrane Separations.

References

E-1.   Graham, A. K. Ed. *Electroplating Engineering Handbook* (New York: van Nostrand, 1971).

E-2.   Mohler, J. B. and W. J. Sedusky. *Electroplating for Metallurgist, Engineer & Chemist*, 3rd ed. (Electrochemical Society, New York: John Wiley & Sons, Inc., 1974).

E-3.   Sittig, M. *Electroplating and Related Metal Finishing—Pollutants and Toxic Materials Control* (Park Ridge, NJ: Noyes Data Corp., 1978).

E-4.   Williams, J. M. and K. B. Keating. "Extended Surface Electrolysis Removes Heavy Metals from Waste Streams," *Chem. Eng.* (February 21, 1983), p. 61.

E-5.   Rice, L. "Heavy Metals Recovery Promises to Pare Water Clean-Up Bills," *Chem. Eng.* (December 1975).

E-6.   Warner, B. "Electrolytic Treatment of Job Shop Metal Finishing Waste Water." EPA 600–2/75–028 (1975).

E-7.   Higley, L. W. Jr. et al. "Lead Dioxide-Plated Titanium Anode for Electrowinning Metals from Acid Solutions." U.S. Bureau of Mines RI 8111 (1976).

E-8.   Dennis, J. K. and T. E. Such. *Nickel and Chromium Plating* (New York: John Wiley & Sons, Inc., 1972).

E-9.   Kemmel, R. and Lieber, H. W. "Electrolytic Recovery of Precious Metals from Dilute Solutions," *J. Metals* 33 (10):45 (1981).

E-10. Cole, E. R. Jr., and T. J. O'Keefe. "Insoluble Anodes for Electrowinning Zinc and Other Metals." U.S. Bureau of Mines RI 8531 (1981).

E-11. Brown, A. P. et al. "The Electrorefining of Copper from Cuprous Ion Complexing Electrolyte," *J. Metals* 33:49 (1981).

E-12. Mussler, R. E. and R. E. Siemens. "Electrowinning Nickel and Cobalt from Domestic Laterite Processing." U.S. Bureau of Mines RI 8604 (1982).

E-13. Bewer, G., H. DeBrodt, and H. Herbst. "Titanium for Electrochemical Processes," *J. Metals* 34:37 (1982).

E-14. Baninati, C. A. and W. J. McLay. "Electrolytic Metal Recovery Comes of Age," *Plating Surf. Finish.* 70:26, (1983).

E-15. Cook, G. M. "The Direct-Electrowinning Process." Op. cit., p. 59.

E-16. Kelsall, G. H. "Fluidized-Bed Electrodes." Op. cit., p. 63.

E-17.   Cushnie, G. C. *Electroplating Wastewater Pollution Control Technology* (Park Ridge, NJ: Noyes Publishing, 1985).

E-18.   *The Canning Handbook on Electroplating* (Birmingham, England: W. Canning Ltd., 1978).

E-19.   Berkowitz, J. B., J. T. Funkhouser, and J. I. Stevens. "Unit Operations for Treatment of Hazardous Industrial Wastes." NTIS PB 275054 and PB 275287 (1978).

E-20.   Pojasek, R. B., Ed. *Toxic and Hazardous Waste Disposal*, Vol. 4 (Ann Arbor, MI: Ann Arbor Science Publishers, Inc., 1980).

E-21.   "Environmental Pollution Control Alternatives: Sludge Handling, Dewatering, and Disposal Alternatives for the Metal Finishing Industry." EPA 625/5-82-018 (1982).

E-22.   Dean, J. G., et al. "Removing Heavy Metals from Waste Water," *Environ. Sci. Technol.* 6:518–522 (1972).

E-23.   Coleman, R. T., et al. "Sources and Treatment of Wastewater in the Nonferrous Metals Industry." EPA-600/2-80-074; NTIS PB80-196918 (1980).

E-24.   Cochran, A. A., et al. "Development and Application of the Waste-Plus-Plus Process for Recovering Metals from Electroplating and Other Wastes." U.S. Bureau of Mines RI 7877 (1974).

E-25.   Crumpler, E. P., Jr. "Management of Metal Finishing Sludges." EPA 530/SW/461; PB 263746 (1977).

E-26.   Bhatia, S. and R. Jump. "Metal Recovery Makes Good Sense!" *Environ. Sci. Technol.* 11:752 (1977).

E-27.   Biswas, A. K. and W. G. Davenport. *Extractive Metallurgy of Copper* (New York: Pergamon Press, 1976).

E-28.   Swank, C. A. BEWT (Water Engineers) Ltd. "Metals Recovery System. Lancy Laboratories: Hazardous Waste Handling," *Met. Finish.* 80:89 (1982).

E-29.   Doroshkevich, A. P., S. V. Karelov, V. I. Rybnikov, I. F. Khudyakov, and L. M. Gryaznukhina. "Behavior of a Steel Electrode During Anodic Polarization in an Ammoniacal Electrolyte," *Izv. Vyssh. Uchebn. Zaved. Tsvetn. Metall.* (5):28–31 (1980).

E-30. Yajima, Y. and S. Toba. "Treatment of Wastewater from Chemical Nickel Plating Processes and Recovery of Nickel," *Jitsumu Hyomen Gijutsu* 29(6):290–298 (1982).

E-31. Ireland, I. R. "Seeding of Electroless Copper Waste for Copper Removal and Recovery as Elemental Copper," in Proceedings IPC Workshop Water Pollution Control, Paper No. 15, 7 pp. (Evanston, IL: Institute for Interconnecting and Packaging Electronic Circuits, 1978).

E-32. Suzuki, A., Y. Umemiya, and S. Iizuka. "Copper Recovery from Copper or Its Alloy Treating Solutions." Japanese Patent 53086627 (1978).

E-33. Oehr, K. H. "Cyanide-Containing Wastewater Treatment." U.S. Patent 4145268 (1979).

E-34. Aslanov, N. N. and N. F. Avetisyan. "Combining Electrolysis and Extraction for the Recovery of Nonferrous Metals," *Mater. Nauchno-Tekh. Soveshch. Kompleksn. Ispol'z. Syr'evykh Resur. Predpr. Tsvetn. Metall.*, Meeting Date 1974, 206–225. Adibekyan, A., Ed. Kh. Izd. "Aiastan": Yerevan, U.S.S.R. (1977).

E-35. Mizumoto, S., H. Nawafune, and M. Kawasaki. "Electrolytic Recovery of Metals from Sludges," *Mem. Konan. Univ. Sci. Ser.* 23:35–44 (1979).

E-36. Bettley, A., A. Tyson, S. A. Cotgreave, and N. A. Hampson. "The Electrochemistry of Nickel in the Chemelec Cell," *Surf. Technol.* 12(1):15–24 (1981).

E-37. Portal, C. and G. M. Cook. "Electrolytic Treatment of Plating Wastes." U.S. Patent 4226685 (1980).

E-38. Kametani, H., M. Kobayashi, T. Mitsuma, and K. Goto. "Wastewater Treatment by Suspension Electrolysis," *Kenkyu Hokokushu—Kinzoku Zairyo Gijutsu Kenkyusho* 3:280–291 (1982).

E-39. Jehanne, Y. and M. Bonneau. "Modular Electrolysis Cell for Metal Recovery." French Patent 2446331 (1980).

E-40. Doroshkevich, A. P., S. V. Karelov, I. F. Khudyakov, A. E. Sokolov, and B. I. Korobitsyn. "Electrochemical Treatment of Bimetallic Wastes," *Tsvetn. Met.* (6):45–8 (1981).

E-41.   Baczek, F. A., B. C. Wojcik, D. M. Lewis, and R. C. Emmett. "The Electroslurry Process for Copper Recovery from Smelter and Refinery Wastes," in *Process Fundam. Consid. Sel. Hydrometall. Syst.*, M. C. Kuhn, Ed. (New York: Society of Mining Engineering AIME, 1981), pp. 125–141.

E-42.   Kreysa, G. "Electrochemical Cell." U.S. Patent 4278521 (1981).

E-43.   Lancy, L. E., W. F. Stevens, and F. L. Sassaman. "Recovery of Metal from Solutions Used in Electroplating and Electroless Plating." Brazilian Patent 8301512 (1983).

E-44.   Spearot, R. M. and J. V. Peck. "Recovery Process for Complexed Copper-Bearing Rinse Waters," *Environ. Prog.* 3(2):124–128 (1984).

E-45.   Cannell, J. F. "Electrolytic Process and Apparatus for the Recovery of Metal Values." European Patent 5007 (1979).

E-46.   Gorodetskii, Y. S., A. M. Romanov, R. V. Drondina, and L. F. Ignatova. "Recovery of Copper from Treated Electroless Copper Plating Solutions," *Deposited Doc.*, VINITI 1017–1019 (1979).

E-47.   Tanaka, T., M. Tanaka, and T. Tanaka. "Recovery of Useful Metals from Waste Sludge in Metal Coating." *Jpn. Kokai Tokkyo Koho*, Japanese Patent 54138801 (1979).

E-48.   Raats, C. M. S. and M. A. Geelen. "Process and Apparatus for Electrolytically Removing Metal Ions from a Dilute Solution Thereof." European Patent 5580 (1979).

E-49.   Mansur, F. and C. Walker. "Plating Waste Treatment and Metals Recovery," *Insul./Circuits* 25(13):33–38 (1979).

E-50.   Burton, S. A. C. "Electrolysis Method and Apparatus for Treating Solutions." U.K. Patent 2028870 (1980).

E-51.   Hebble, T. L., V. R. Miller, and D. L. Paulson. "Recovery of Principal Metal Values from Electrolytic Zinc Waste." U.S. Bureau of Mines RI 8582 (1981).

E-52.   Tomlinson, D. "Electrolytic Recovery of Metals. An Inert Fluidized Bed Process for Recovery from Dilute Solutions," *Riv. Mecc.* 753:83–85 (1982).

E-53.   Abraham, Z., B. Kobor, R. Vandor, T. Stollar, and F. Mohacsi. "Recovery of Copper by Electrolysis from Copper Tetramine Chloride Based Spent Etching Solutions for Copper Foils." Hungarian Patent 21295 (1981).

E-54.   Miller, V. R. and D. L. Paulson. "Recovering Accessory Minerals from Lead and Zinc Process Wastes," *Resour. Conserv.* 9:95–104 (1982).

E-55.   Leclerc, G. "Application of Electrolysis to Recovery [of Metals] from Solutions", *Tech. Mod.* 74(1-2): 91–93 (1982).

E-56.   Brent, D. "Recovery of Metals from Industrial Wastes," *Chem. Prod.* 11(2):45 (1982).

E-57.   Tison, R. P. "Packed and Tumbled-Bed Electrochemical Reactors for Metals Recovery," *Environ. Prog.* 2(1):70–74 (1983).

E-58.   Fischer, G. "Metal Recovery by Electrolysis," *Galvanotechnik.* 74(2):145–150 (1983).

E-59.   Nippon Mining Co., Ltd. "Electrolytic Copper Removal from Solution." *Jpn. Kokai Tokkyo Koho*, Japanese Patent 57185996 (1982).

E-60.   Konicek, M. G. and G. F. Platek. "Reticulate Electrode Cell Removes Heavy Metals from Rinse Waters," *New Mater. New Proc.* 2:232–235 (1983).

E-61.   Lieber, H. W. "Development of a Process for the Electrolytic Recovery of Metals from Diluted Wastewater," *Forschungsber. — Bundesminist. Forsch. Technol., Technol. Forsch. Entwickl.* BMFT-FB-T 82-224 (1982).

E-62.   Diamond Shamrock Corp. "Metal Recovery from Electroplating Wastewater." *Jpn. Kokai Tokkyo Koho*, Japanese Patent 58027687 (1983).

E-63. Bishop, P. L. and R. A. Breton. "Electrolytic Recovery of Copper from Chelated Waste Streams," in *Toxic Hazardous Waste, Proceedings of the Mid-Atlantic Industrial Waste Conference*, 15th, M. D. LaGrega and L. K. Hendrian, Eds. (Boston: Butterworth Publishers, Inc., 1983), pp. 584–596.

E-64. Samhaber, W. "Electrochemical Separation of Heavy Metals [from Waste Waters]," *Chem.-Ing.-Tech.* 56(3):246–247 (1984).

E-65. Warheit, K. E. "The Recovery and Treatment of Metals from Spent Electroless Processing Solutions," *Annual Technical Conference Proceedings — American Electroplating Society*, 70th, C-4 (1983).

E-66. Ruml, V., M. Soukup, and P. Zoltan. "Recovery of Copper, Nickel, Zinc, and Cadmium from Waste Sludges." Czechoslovakian Patent 214012 (1984).

E-67. Soukup, M., V. Ruml, and J. Tenygl. "Bipolar Rotary Electrolyzer for Electrolysis of Dilute or Low-Conducting Solutions." Czechoslovakian Patent 206884 (1984).

E-68. Simonsson, D. "A Flow-by Packed-Bed Electrode for Removal of Metal Ions from Wastewaters," *J. Appl. Electrochem.* 14(5):595–604 (1984).

E-69. Lee, J. K. and H. S. Chun. "Electrolytic Recovery of Metals from the Plating Rinse Water with Fluidized Bed Electrode Reactor," *Kumsok Pyomyon Choli.* 17(1):1–6 (1984).

E-70. Menyhart, J., M. Fisch, and J. Fulop. "Recovery of Metal Compounds from Metal Oxide-Containing Wastes in Electroplating." Hungarian Patent 32167 (1984).

E-71. Baerring, N. E. "Recycling of Nickel-Cadmium Batteries and Process Wastes — Processes and Operations of the New SAB NIFE Plant," Ed. Proc. — Int. Cadmium Conf., 4th, D. Wilson and R. A. Volpe, Eds. Cadmium Association, London (1983), pp. 58–60.

E-72. Wiegand, V. "Treatment and Recovery of Cadmium from Steelworks Flue Dust," Ed. Proc. — Int. Cadmium Conf., 4th, D. Wilson and R. A. Volpe, Eds. Cadmium Association, London (1983), pp. 67–69.

E-73.   Kohl, D., J. Walther, J. Schmidt, H. Kotschy, and H. J. Hasenstein. "Separation of Copper from Copper-Containing Solutions." German Patent 212950 (1984).

E-74.   Doroshkevich, A. P., V. F. Bogdashev, S. V. Karelov, and B. I. Korobitsyn. "Processing of Bimetallic Wastes with Production of Copper-Zinc Powders," *Tsvetn. Met. (Moscow)* (1):14–17 (1985).

E-75.   Apel, M. L., P. S. Fair, and J. P. Adams. "Design and Application of a Spray Rinsing System for Recycle of Process Waters." Report, EPA/600/D-84/246 ; Order No. PB85–106722/GAR (1984).

E-76.   Machnadz, E. "Industrial Wastes in a Processing Plant of Nonferrous Metals," *Rudy Met. Niezelaz.* 29(8):363–366 (1984).

E-77.   Toppan Printing Co., Ltd. "Metal Recovery from Industrial Wastewater," *Jpn. Kokai Tokkyo Koho,* Japanese Patent 60077937 (1985).

E-78.   Dattilo, M., E. R. Cole, and T. J. O'Keefe. "Recycling of Zinc Waste for Electrogalvanizing," *Conserv. Recycl.* 8(3–4):399 (1985).

E-79.   Prochazka, J., M. Mrnka, J. Formanek, J. Sladkovska, A. Heyberger, V. Bizek, P. Javorek, and V. Sychra. "Recovering Molybdenum and Copper from Wastewater in Producing Phthalocyanine Pigments." Czechoslovakian Patent 218293 (1985).

E-80.   Farkas, J. and G. D. Mitchell. "An Electochemical Treatment Process for Heavy Metal Recovery from Wastewaters," *AIChE Symp. Ser.* 81(243, *Sep. Heavy Met. Other Trace Contam.*):57 (1985).

E-81.   Cruceru, M., C. Licaret, and S. Duda. "Recovery of Copper from Electrotechnical Industry Wastes." Romanian Patent 86772 (1985).

E-82.   Kovalev, V. V., M. I. Sudvarg, M. G. Zhurba, and O. V. Kovaleva. "Apparatus for Extracting Nickel from Wash Water." U.S.S.R. Patent 1203123 (1986).

E-83.   Horibe, K. "Recovery of Copper from Wastewater." *Jpn. Kokai Tokkyo Koho*, Japanese Patent 61009530 (1986).

E-84.   Jegrand, J., J. M. Marracino, and F. Coeuret. "Copper Recovery from Dilute Solutions in the Falling-Film Cell," *J. Appl. Electrochem.* 16(3):365–373 (1986).

E-85.   Edson, G. I. "Electrolytic Reactor." U.S. Patent 4585539 (1986).

E-86.   Shen, N., L. Wu, and M. Yang. "Palladium-Nickel Alloy Plating—A Substitute for Gold Plates on Printed Circuit Boards," *Diandu Yu Huanbao* 5(1):5–8 (1985).

E-87.   Nakaji, Y. and J. Oishi. "Method and Apparatus for Metal Recovery." *Jpn. Kokai Tokkyo Koho*, Japanese Patent 61106788 (1986).

E-88.   Nakaji, Y. and J. Oishi. "Apparatus for Metal Recovery." *Jpn. Kokai Tokkyo Koho*, Japanese Patent 61104096 (1986).

E-89.   Nagashima, S. and K. Higashi. "Recovery of Nickel from Wastes of Electroplating by Electrolysis With and Without a Separating Diaphragm," *Kenkyu Hokoku—Tokyo-toritsu Kogyo Gihutsu Senta* (14):192 (1985).

E-90.   Higashi, K. and S. Nagashima. "Recovery of Nickel from Plating Rinse Water by Electrolysis," *Kenkyu Hokoku—Tokyo-toritsu Kogyo Gijutsu Senta* (14):154–158 (1985).

E-91.   Andreas, B., I. Pulpit, K. Merck, and D. Vetter. "Electrolytic Removal of Copper and Cyanide from Solutions Containing Alkali and Copper Cyanide." German Patent 230566 (1985).

E-92.   Hirano, K. and T. Tanabe. "Experiments on Continuous Electrolytic Copper Recovery from Rinse Water of Copper Electroplating," *Kenkyu Hokoku—Kanagawaken Kogai Senta* 8:13–17 (1986).

E-93.   Yasuda, M. and S. Yasukawa. "Electrolytic Treatment of Cyanide Wastes. III. Effect of the Addition of Sodium Chloride on the Anodic Oxidation of the

Copper(I) Cyano Complex," *Denki Kagaku Oyobi Kogyo Butsuri Kagaku* 54(2):149–152 (1986).

E-94. Wahl, K. L. and F. Reinhard. "Avoidance of Hydroxide Sludges in Electroplating by Total Metal Recovery with the Extended RMA System," *Galvanotechnik* 77(10):2421–2424 (1986).

E-95. Gana, R. E., M. G. Figueroa, and A. A. Parodi. "Operation of an Industrial Pilot Plant for Copper Electrorefining with the Support-Anode System," Min. Lat. Am., Pap. Min. Lat. Am./Min. Latinoam, Conf. (1986), pp. 131–134.

E-96. Aoki, H., T. Hiroshi, and E. Nishimura. "Recovery of Metals from Spent Dry Batteries." *Jpn. Kokai Tokkyo Koho*, Japanese Patent 61261443 (1986).

E-97. Farrar, L. S. "Recovery of Palladium from Plating Operations," *Plating Surf. Finish.* 74(3):60–61 (1987).

E-98. Huggins, R. G. and A. D. Towery. "Waste Minimization. A Case Study," *Hazard. Waste Hazard. Mater.* 4(1):43–45 (1987).

E-99. Kermer, K. and D. Wunderlich. "Recovery of Copper from Spent Sulfuric Acid-Acetic Acid-Hydrogen Peroxide Etching with Simultaneous Regeneration of Etchant." East German Patent 237332 (1986).

E-100. Dou, Z., Q. Song, X. Li, M. Lin, W. Lan, and Y. Zhao. "Recovery of Nickel and Other Metals from Nickel Alloy Scrap," *Conserv. Recycl.* 10(1):21–26 (1987).

E-101. Bin, W., X. Chen, Q. Chen, and Y. Lu. "Comprehensive Utilization of Residues Produced in the Hydrometallurgical Processing of Copper Anode Slimes," *Kuangye Gongcheng* 6(4):46–49 (1986).

E-102. Gutt, G. "Galvanic Cell for the Recovery of Metals from Electroplating Waste Water." Romanian Patent 90511 (1987).

E-103. Holly, J. and P. Pauliny. "Apparatus and Method for Recovery of Copper from Dross by Electrolysis." Czechoslovakian Patent 234328 (1987).

E-104. Korecki, T., W. Kamper, and H. Szalapski. "Method of Electrochemical Neutralization of Solid Postelectroplating Wastes, Especially Wastes Forming During Copper Plating Using Potassium Cyanide." Polish Patent 131796 (1986).

E-105. Avci, E. "Electrolytic Recovery of Copper from Dilute Solutions Considering Environmental Measures," *J. Appl. Electrochem.* 18(2):288–291 (1988).

E-106. Takazawa, Y. and K. Narita. "Recovery of Tin from Copper Scrap by Leaching and Electrolysis." *Jpn. Kokai Tokkyo Koho*, Japanee Patent 63060241 (1988).

E-107. Guelbas, M. "Sewage Water and Recycling Technique. Valuable Material Recovery by Electrolysis," *Metalloberflaeche* 42(4):191–195 (1988).

E-108. Ishisaki, C. "A Method and Apparatus for Regenerating a Copper Chloride Etching Liquid Waste." *Jpn. Kokai Tokkyo Koho*, Japanese Patent 62297476 (1987).

E-109. Sakata, A. and K. Suzuki. "Nickel Recovery from Waste Metal." *Jpn. Kokai Tokkyo Koho*, Japanese Patent 63083234 (1988).

E-110. Takazawa, Y., Y. Koizumi, and H. Okamoto. "Metal Value Recovery from Soldered or Tinplated Metals." *Jpn. Kokai Tokkyo Koho*, Japanese Patent 63157821 (1988).

E-111. Tian, B. and X. Zhang. "Recovery of Copper from Wastewater by Electrolysis," *Diandu Yu Huanbao* 8(3):22–23 (1988).

E-112. Nanjo, M., Y. Ito, Y. Sato, and T. Sato. "Fundamental Studies of Municipal Waste Treatment and Utilization (VII). Copper-Elimination Step of Recycling Process for Niobium in Niobium-Titanium and Niobium Stannide (Nb_3Sn) Superconductor Wire Scrap," *Tohoku Daigaku Senko Seiren Kenkyusho Iho*, 43(2):215–226 (1987).

E-113. Sun, X. and F. Yin. "Separation and Recovery of Nickel and Cadmium from Wastes by Fluidized-Bed Electrolysis," *Shanghai Huanjing Kexue* 7(3):23–24 (1988).

E-114. Wiaux, J.P., "Recycling in Electroplating Plants. Treatment of Wastewater by Electrolysis," *Oberflaeche-Surf.* 29(10):16, 18, 20–21 (1988).

E-115. Avci, E. "Electrolytic Recovery of Nickel from Dilute Solutions," *Sep. Sci. Technol.* 24(2):317–324 (1989).

E-116. Anon. "Electrolytic Recovery of Heavy Metals: Nickel and Copper in Chemical Baths," *Galvano-Organo-Trait. Surf.* 57(589):786–790 (1988).

E-117. Sun, X. and F. Yin. "Use of Fluidized-Bed Electrolysis to Separate and Recover Copper and Zinc from Copper Washing Wastewater," *Huanjing Wuran Yu Fangzhi* 10(3):23–25 (1988).

E-118. Khudyakov, I. F., S. V. Mamyachenkov, and S. V. Karelov. "Electrochemical Separation of Copper and Tin from Recycled Copper-Base Alloys," *Tsvetn. Met. (Moscow)* (4):44–46 (1989).

E-119. Fischer, G. "Electrodialysis for Concentrating and Recovering Metal Salt Solutions," *Metalloberflaeche* 43(7):309–313 (1989).

E-120. Kastening, B. "Technique for Integrated Copper Recovery in Etching of Printed-Circuit Boards," *DECHEMA-Monogr.*, 117(*Elektrochem. Elektron.*): 115–127 (1989).

E-121. Herbst, R. J. and R. R. Renk. "Electrolytic System for Wastewater Treatment." U.S. Patent 4872959 (1989).

E-122. Nedelec, G. and J. J. Lasvaladas. "Removal of Metals, Especially Copper, from Wastewaters by Electrolysis." French Patent 2629446 (1989).

E-123. Sabatini, J. S., E. L. Field, E. S. Shanley, and D. A. Weiler. "Electrolytic Technology in the Metals Industry: A Scoping Study: Final Report." Report, EPRI-EM-6098; from: *Energy Res. Abstr. 1989*, 14(9): Abstr. No. 17944 (1989).

E-124. Boyd, D. M. and R. J. Fulk. "Treatment of Plating Wastewater Without Sludge," *Proceedings of the 43rd Purdue Industrial Waste Conference*, Volume Date 1988 (Chelsea, MI: Lewis Publishers, Inc., 1989), pp. 499–504.

E-125. Ogawa, K. "Recovery of Metals from Waste Solutions." *Jpn. Kokai Tokkyo Koho*, Japanese Patent 54071772 (1979).

E-126. Mitzner, R., A. Rosenthal, and C. Niemann. "Determination of Pollutant Concentrations in the Wastewaters from the Berlin Metalworks and Semifinished Metal Products Works (BMHW) and Optimization of Parameters for the Electrolytic Recovery of Copper from the Etching Solutions of This Plant," *Nachr. Mensch-Umwelt* 7(1-2, Pt. 2):92-94 (1979).

E-127. Brunschweiler, A. and H. Maurer. "Metal Recovering Cell for the Purification of Industrial Process Solutions and Wastewaters." German Patent EP 36640 (1981).

E-128. Zaidan Hojin Kagaku Gijutsu Shinkokai. "Copper Recovery by Electrolysis from Chloride Etching Solution." *Jpn. Kokai Tokkyo Koho*, Japanese Patent 55145175 (1980).

E-129. Nazarova, G. N., L. V. Kostina, M. D. Venkova, N. V. Kovaleva, and V. A. Ravcheev. "Use of Electrochemical Technology in the Separation of Nickel from Industrial Solutions at Hydrometallurgical Plants," *Komb. Metody Pererab. Medno-Nikelevykh. Rud* 113-119 (1979).

E-130. Robertson, P. M., J. Leudolph, and H. Maurer. "Improvements in Rinsewater Treatment by Electrolysis," *Plating Surf. Finish.* 70(10):48-52 (1983).

E-131. Abe, T. and T. Kato. "Acid Solution Production by Electrolysis in Copper Refining." *Jpn. Kokai Tokkyo Koho*, Japanese Patent 60245795 (1985).

E-132. Stewart, T. L. and J. N. Hartley. "Electrolytic Recovery of Copper and Regeneration of Nitric Acid from a Copper Strip Solution," *Energy Reduct. Tech. Met. Electrochemical Processes, Proceedings Symposium* (Metallurgical Society: Warrendale, PA, 1985), pp. 13-25.

E-133. Shilin, A. I., L. D. Sedova, and L. I. Ruzin. "Recovery of Nonferrous and Precious Metals in Electroplating," *Ekon. Tekhnol. Gal'van. Pr-va, M.* 125-127 (1986).

E-134. Wei, T. Y., I. S. Shaw, and Y. C. Hoh. "Electrolytic Recovery of Low Concentration Cupric Ions by a Fluidized-Bed Electrode Reactor," *J. Chin. Inst. Chem. Eng.* 18(2):83–91, (1987).

E-135. Kiessling, R., V. Pizak, and H. Wendt. "Recovery of Lead from Used Storage Batteries and Reduction Plate Therefore." German Patent 3402338 (1985).

E-136. Mowla, D., H. Olive, and G. Lacoste. "Application of Volumetric Electrodes to the Recuperation of Metals in Industrial Effluents. III. Potential Distribution and Design of Radial Field Electrodes," *Electrochim. Acta* 28(6):839–846 (1983).

5.4 ION EXCHANGE

Ion exchange has received considerable attention for separation and concentration of a variety of metals from waste effluents[I-1–I-10]. Extensive work has been done in developing removal and, to a lesser extent, recovery of metals such as chromium, cobalt, copper, cadmium, nickel, iron, and zinc from pickling acids and plating wastes. Strong cation exchange resins, usually with sulfonic acid exchange sites on a cross-linked polystyrene resin, are most commonly used for soluble metal cations with acid resin regeneration. Anion exchange resins with various amine base functionalities for anion exchange sites incorporated into cross-linked polymer matrices are used for separation of metals in anionic form, notably chromate/dichromate, with resin regeneration with an alkali hydroxide. This technology is highly developed and is available for commercially designed systems.

Promising recent progress for cationic exchange separation of the metals, such as cadmium, cobalt, copper, nickel, zinc, etc., has been obtained with a variety of chelate exchange functionality such as pyridyl imidazole,[I-43] polyethylene imine dithiocarbamate,[I-46] bispicolylame or N-(2-hydroxypropyl) picolylamine,[1-67] iminodiacetic-polystyrene,[1-68] imino diacetic/polyacrylamide,[1-89] N,N,N',N[1],tetrakis (2-hydroxypropyl) ethylene diamine,[1-91] copolymer divinyl benzene with primary or secondary phosphinic acid,[1-93] polystyrene + dehydrodithizone copolymer [1-94] and bispicolylamine[1-96].

A summary of the principal types of ion exchange resins used for separation of metals such as Cd, Co, Cr, Cu, Hg, Mn, Ni, Pb, Zn, and precious metals is given in Table 5.5.

Recent efforts have been directed to designing metal removal from metal finishing and electroplating waste effluents with modular systems involving centralized treatment and metal recovery plants integrated with widely dispersed, small-scale collection systems at individual waste generation sites[1-106, 1-107].

There are some less conventional ion exchange systems applicable to separation and concentration of soluble metals. One such system involves metal exchange with polyelectrolytes such as polygalacturonic acid or polymethacrylic acid,[1-104] using acid regeneration to recover the metal. Another type of system consists of spun fibers (polyamines, polyvinyl pyridine, polystyrene, polymethacrylate, etc.) with cation exchange sites introduced by sulfonation or anion exchange sites introduced by inclusion of basic organic (quaternary amine) functionality into the polymer.[1-105] Disposable ion exchange systems such as vermiculite have been proposed[1-16] for the nonregenerable removal without recovery of metals from wastewaters.

Ion exchange metal removal from waste effluents is most appropriately applied to relatively dilute systems $< \approx 1000$ ppm metal content and to systems with minimal amounts of competing cations. Selectivity for multicomponent systems is only attainable at the expense of more sophisticated process design and higher plant costs. Waste systems treated must be relatively free of insoluble colloidal materials to maximize exchange efficiency and regenerability of the resin.

Table 5.5 Metal Recovery by Ion Exchange

Waste system	Metals	Ion exchange agent	Chemical type	Metal separation efficiency %	Ref.
Metal wastewater	Ni	Uniselec UR 20	N. type chelate resin	Not cited	I-17
Waste ammon. liquors	Ni	Zerolit Z236	Carboxylic acid resin	Not cited	I-20
Wastewater	Cu, Ni	Coal-based chelating agent		Not cited	I-21
Galvanic hydroxide waste	Cr^{6+}	Anion exchange resin		Not cited	I-23
Plating waste	Ag, Cu	Lewatat, MP 64 H, Lewatit, SP112 Lewatit MP500	Anonic cationic hydroxide	Not cited	I-24
Flue gas wastewater	Ni, V	L Uniselec UR 50	Carboxylic acid resin	94 Ni, 92 V	I-28
Cu wastewaters	Cu	Sindex-C26	Cationic resin	97–98 Cu	I-29
Metal wastewaters	Heavy metals	Zerolit 225	Cationic	98–99	I-32
Plating waste	Ni	Dowex 50W X 8	Cationic	~99 Ni	I-33
Pb smelting wastes	Cu, Zn	AV–17–10P	Macro porous resin	98–99 Cu; 97 Zn, Au; 60–75 Ag	I-35
Metal wastewater	Ni	Woltatit CA-2U	Cationic	>80 Ni	I-39
Plating wastes	Ni, Rh		Cationic resin	Not cited	I-36
Plating wastes	Cu, Ni		Chelate resin	Not cited	I-41
Plating wastes	Cd, Cu, Zn		Anionic resin	Not cited	I-42
Metal wastes	Cu	XF4195, F4196	Pyridyl imidazole	Not cited	I-43

Source	Metal	Resin/Product	Description	Recovery	Ref.
Metal wastes	Cu		Polyethyleneimine dithiocarbamate	99.5 Cu.	I-46
Cu plating waste	Cu	D-231	Dialkylamine sulfate	98 Cu	I-47
Semiconductor wastes	Cu	Chelex-100	Chelate resin	>90 Cu	I-49
Plating wastes	Au, Cu		Polyolefin fiber	Not cited	I-54
Plating wastes	Ag, Au, Cr, Cu, Ni, Zn	Vionit	Cationic		I-56
Plating wastes	Zn	TRA 400			I-58
Metal wastes	Cu, Ni, Zn	Amberlite 120	Sulfonic acid-divinyl Benzene		I-61
Plating wastes	Cu	Amberlite IR120, IRC84	Cationic		I-66
Electroless plating waste	Cu		bis-Picolylamine or N-(2-hydroxypropyl) picolylamine		I-67
Metal wastes	Co, Cu, Ni, U		Polyethyleneimine resin		I-69
Cat. mfr. waste	Cu, Zn	KMD, Vofatit-KPS	Cationic		I-74
Photo wastewater	Ag, Ni		Naclinoptilolite		I-75
Cu leach solutions	Co	MCIX	Not specified	92 Co	I-78
Plating waste	Cu		Waste rennet-casein plastic		I-81
Metal wastes	Cu		Iminodiacetic/poly acrylamide		I-89

Table 5.5 continued

Waste system	Metals	Ion exchange agent	Chemical type	Metal separation efficiency %	Ref.
Sugar processing waste	Various metals		Cationic		I-90
Electroless plating	Cu, Ni		Anionic Quatenary/polystyrene N, N, N', N', tetrakis (2-hydroxypropyl) ethylene diamine		I-91
Metals in solution	Ag, Am, Cu, Eu, Hg, Th, Zn		Copolymer divinyl-benzene/primary, secondary phosphinic acid		I-93
Metals in Cl. solution	Au, Ir, Os, Pd, Pt, Ru		Polystyrene/de-hydrodithizone copolymer		I-94
Metals in solution	Co, Fe, Ni, Pb, Zn	Dowex XPS4195	Chelate resin Bispicolylamine		I-96
Plating solutions	Cd, Cu, Fe, Cr(III), Pb, Za	Duolite ES-467 Dowex XFS4195, Amberlite IRC-718	Chelate resins		I-97
Metals in Cl solution	Co, Cu Ni, Zn		Iminodiacetic resin		I-98
Cu waste	Cu	Wolfatite KPS	Electrodialysis with anionic/cationic membranes		I-101

References

I-1. Costa, R. L. "Purification of Chromic Acid Solutions with Cation Exchange Resins," *Ind. Eng. Chem.* 42:308 (1950).

I-2. Reents, A. C. and D. M. Strongquist. "Removal of Chromic Acid from Dilute Rinse Waters with Anion Exchange Resins," *Purdue Univ. Eng. Bull. Ext. Ser. No. 79* 36:462 (1952).

I-3. Keating, R. J., R. Doorin, and V. J. Calise. "Recovery of Chromic Acid from Chrome Plating and Anodizing Rinses," *Proceedings of the 9th Purdue Industrial Waste Conference* (West Lafayette, IN: Purdue University, 1955), p. 1.

I-4. Bueltman, C. G. "Ion Exchange Treatment of Industrial Wastes," *Sewage Ind. Wastes* 29:1018 (1957).

I-5. Kraus, K. A., D. C. Michelson, and F. Nelson. "Adsorption of Negatively Charged Complexes by Cation Exchangers," *J. Am. Chem. Soc.* 81:3204–3207 (1959).

I-6. Abrams, I. M. "Ion Exchange Applications in Electroplating," *7th Far Western Regional Technical Forum*, American Electroplaters' Society, Santa Clara Valley Branch, Burlingame, CA (1967).

I-7. Smithson, G. F. "An Investigation of Techniques for Removal of Chromium from Electroplating Wastes," *Battelle Memorial Inst. EPA Water Pollution Contract Research Series* (1971).

I-8. Miller, W. S. and A. B. Mindlerdler. "Ion Exchange Separation of Metal Ions from Water and Waste Waters," in *Recent Developments in Separation Science*, Vol. III A (Cleveland: CRC Press, 1977).

I-9. Ramon, R. and E. L. Karlson. "Reclamation of Chromic Acid Using Continuous Ion Exchange," *Plating Surf. Finish.*, 64(6):40–42 (1977).

I-10. Skovronek, H. S. and M. K. Stinson. "Advanced Treatment Approaches for Metal Finishing Wastewater (Part 2)," *Plating Surf. Finish.* 64, 11, 24 (1977).

I-11. Hall, E. P., D. J. Lizdas, and E. E. Auerbach. "Recovery Techniques in Electroplating," *Plating Surf. Finish.* 66, 2, 49 (1979).

I-12. Jones, K. C. and R. A. Pyper. "Copper Recovery from Acidic Leach Liquors by Continuous Ion Exchange and Electrowinning," *J. Metals* 31:19 (1979).

I-13. Calmon, C. and G. P. Simon. *Ion Exchangers for Pollution Control* (Boca Raton, FL: CRC Press, Inc., 1979).

I-14. Gold, H. and C. Calmon. "Ion Exchange: Present Status, Needs and Trends," *AIChE Symp. Ser. 192* Vol. 76, 60 (1980).

I-15. Kennedy, D. C. "Predict Sorption of Metals on Ion-Exchange Resins," *Chem. Eng.* (1980), p. 106.

I-16. Keramida, J. and J. E. Etzel. "Treatment of Metal Plating Waste Water with Disposable Ion Exchange Material," *Proceedings of the 37th Purdue Industrial Waste Conference,* J. B. Bell, Ed. (Ann Arbor, MI: Ann Arbor Science, 1983).

I-17. Shinmura, T., H. Ueshima, T. Shibata, and T. Otsuki. "Recovery of Nickel in Waste Water." *Jpn. Kokai Tokkyo Koho*, Japanese Patent 53109894 (1978).

I-18. Hayashi, T., H. Fukuoka, A. Kameyama, and Y. Tanaka. "Recovery of Useful Metals from Wastewater Sludge." *Jpn. Kokai Tokkyo Koho*, Japanese Patent 53130855 (1978).

I-19. Lazzara, M. "Copper and Nickel Recovery in Rinse Waters Arising from Copper- and Acidic Nickel-Plating Lines in the Electroplating Industry," *Inquinamento* 20(9):193, 195, 197, 199, 201 (1978).

I-20. Price, M. J., S. Cutfield, and J. G. Reid. "Nickel Recovery from Nickel Refinery Waste Liquor." *North Queensland Conference, Australasian Inst. Min. Metall.* (Parkville, Aust.: Australasian Inst. Min. Metall., 1978), pp. 355–360.

I-21. Knocke, W. R., T. Clevenger, M. M. Ghosh, and J. T. Novak. "Recovery of Metals from Electroplating Wastes." *Proceedings of the 33rd Purdue Industrial Waste*

Conference (Ann Arbor, MI: Ann Arbor Science, 1978), pp. 415–426.

I-22.   Watanuki, M. "Recovery of Nickel and Copper from Wastewater." *Jpn. Kokai Tokkyo Koho*, Japanese Patent 53141106 (1978).

I-23.   Mueller, W. and L. Witzke. "Processing Nonferrous Metal Hydroxide Sludge Wastes." U.S. Patent 4151257 (1979).

I-24.   Yabe, E. "Treatment of Plating Waste Solutions." *Jpn. Kokai Tokkyo Koho*, Japanese Patent 54075159 (1979).

I-25.   Van Wijk, H. F. and B. Van Engelenburg. "Recovery of Heavy Metals in Chemical Wastes by Selective Ion Exchange," *PT-Procestech.* 34(5):283–286 (1979).

I-26.   Pelikan, J., J. Slepicka, F. Nekvasil, and F. Pospichal. "Ion Exchange Recovery of Nickel from Rinsing Water," *Povrchove Upravy* 19(1):31–34 (1979).

I-27.   Vaillagou, P. "Industrial Application of Copper Recovery on a Chain of Treatments of Printed Wires," *Trait. Surf.* 169:30–32 (1979).

I-28.   Inoue, T., M. Matsuoka, K. Naito, and T. Shibata. "Recovery of Valuable Metals from Waste Materials in Flue Gas Treatment." *Jpn. Kokai Tokkyo Koho*, Japanese Patent 54104651 (1979).

I-29.   Maji, B. K., V. K. Seth, and B. K. Dutta. "Application of an Ion Exchange Technique in Effluent Treatment: Recovery of Copper from Copper-Bearing Effluents," *Fert. Technol.* 15(3–4):288–290 (1978).

I-30.   Pospichal, F. and R. Styblo. "Ion Exchange Recovery of Nickel from Wastewater. II," *Povrchove Upravy* 19(3):24–27 (1979).

I-31.   Price, K. and C. Novotny. "Water Recycling and Nickel Recovery Using Ion Exchange." U.S. Environmental Protection Agency, Off. Res. Dev., [Rep.] EPA, EPA/600/8–79/014, Conf. Adv. Pollut. Control Met. Finish. Ind., 2nd; PB-297 453 85–7 (1979).

I-32.   Raiter, R. "Regeneration of Strong Cation Exchangers." South African Patent ZA 7805885 (1980).

I-33.   Reynolds, B. A., J. C. Metsa, and M. E. Mullins. "Distribution Ratios on Dowex 50W Resins of Metal Leached in the Caron Nickel Recovery Process." Report, ORNL/MIT-301; from: *Energy Res. Abstr.* 1980, 5(16), (1980).

I-34.   Kholmogorov, A. G., V. P. Kirillova, and S. N. Il'ichev. "Sorbent for the Extraction of Nickel from Solutions," *Tsvetn. Met.* (2):29–30 (1981).

I-35.   Vasil'ev, B. F. "Use of a Macroporous Ion Exchanger for the Detoxication of Cyanide-Containing Wastewater," *Tsvetn. Met.* (1):20 (1981).

I-36.   Suzuki, T., T. Hatsushika, Y. Hayakawa, N. Ayuzawa, and Y. Matsumura. "Reclamation of Rhodium and Nickel Ions from Plating Wastewater by an Ion-Exchange Method," *Conserv. Recycl. 4(4):239–243 (1981).*

I-37.   Mal'tsev, G. I., V. V. Sviridov, Yu. B. Kholmanskikh, N. K. Sitnikova, and B. K. Radionov. "Separation of Indium from Converter Dusts and Sublimates of Copper-Smelting Production," *Kompleksn. Ispol'z. Miner. Syr'ya* (3)47–50 (1982).

I-38.   Goncharova, N. A., I. M. Strukova, E. G. Smirnova, G. M. Mubarakshin, and L.V. Emets. "Sorption of Copper by Different Types of Ion-Exchange Fibrous Materials," *Zh. Prikl. Khim.* (Leningrad) 55(9):2095–2097 (1982).

I-39.   Halle, K., K. Fischwasser, and B. Fenk. "Recovery of Metals from Electroplating Wastes," *Tech. Umweltschutz* 25:120–132 (1982).

I-40.   Donaruma, L. G., S. Kitoh, G. Walsworth, J. K. Edzwald, J. V. Depinto, M. J. Maslyn, and R. A. Niles. "Copper Containing Poly[thiosemicarbazides]," *Polym. Prepr. Am. Chem. Soc. Div. Polym. Chem.: 22(1):149–150 (1981).*

I-41.   Courduvelis, C., G. Gallagher, and B. Whalen. "A New Treatment for Wastewater Containing Metal Complexes," *Plating. Surf. Finish.* 70(3):70–73 (1983).

I-42.   Hamil, H. F. "Removal of Toxic Metals in Electroplating Wash Water by a Donnan Dialysis Process." Report, Order No. PB83–148155. From: *Gov. Rep. Announce. Index (U.S.)* 1983, 83(8):1673 (1982).

I-43.   Green, B. R. and R. D. Hancock. "Useful Resins for the Selective Extraction of Copper, Nickel, and Cobalt," *J. S. Afr. Inst. Min. Metall.* 82(10):303–307 (1982).

I-44.   Roberts, G. M., J. D. Nolan, K. A. Korinek, and L. A. Knerr. "Ion Exchange and Electrochemistry: A Combination of Techniques," *Hydrometall. 81, Proceedings of Society of Chemical Industry Symp.* G6/1-G6/10 (1981).

I-45.   Kauczor, H. W. "New Methods of Metal Separation with Ion Exchange Resins," *Erzmetall* 36(10):474–478 (1983).

I-46.   Hartley, F. R., J. H. Barnes, and C. Bates. "Metal Ion Extraction." European Patent 90551 (1983).

I-47.   Luo, Y. "Recovery of Copper from Copper-Plating Wastewater with Moving-Bed Ion-Exchange Resin," *Huaxue Shijie* 23(7):215–217 (1982).

I-48.   Pimenov, V. B., A. I. Zuev, and V. N. Startsev. "Sorption of Metals from Cyanide Wastewaters of Beneficiation Plants," *Tsvetn. Met.* (12):79–81 (1983).

I-49.   Haas, C. N. and V. Tare. "Application of Ion Exchangers to Recovery of Metals from Semiconductor Wastes," *React. Polym., Ion Exch., Sorbents* 2(1–2):61–70 (1984).

I-50.   Fischwasser, K. and J. Kaeding. "Recovery of Useful Material from Industrial Wastewaters by Ion Exchangers," *Acta Hydrochim. Hydrobiol.* 12(2):183–202 (1984).

I-51.   Manczak, M. "Treatment of Pickling Plant Wastewater." Polish Patent 123825 (1984).

I-52.   Kalalova, E. and Z. Kovarova. "Use of Ostion KS Ion Exchanger for Regeneration of Catalytically Active Metals from Synthetic Diamond Production," *Chem. Prum.* 34(5):241–243 (1984).

I-53. Tautz, B. "Copper Recovery from Ammoniacal Spinning Wastewater of the Copper Oxide Ammonia Cellulose Industry by a Cation Exchanger." East German Patent 200854 (1983).

I-54. Zapisek, S. "A New Ion-Exchange Material for Metal Finishers," *Plating Surf. Finish.* 71(4):34–36 (1984).

I-55. Gladkov, S. Y., N. A. Churilova, N. I. Oklobystin, A. A. Lobachev, E. M. Pakhomova, N. I. Smirnova, S. B. Malarova, and M. I. Kislova. "Recovery of Nickel from Electroplating Wash Waters." U.S.S.R. Patent SU 118707 (1984).

I-56. Patrascu, R. and M. Dumitrescu. "Use of Romanian Ion Exchangers for Recovering Metals from Electroplating Wastewater," *Constr. Masini. 36(2–3):102–107 (1984).*

I-57. Yamasaki, K., T. Morikawa, and S. Eguchi. "Removal and Recovery of Nickel in Wastewater from Chemical Plating by Ion Exchange Method," *Osaki-Furitsu Kogyo Gijutsu Kenkyusho Hokoku* (84):10–15 (1984).

I-58. Gupta, A., E. F. Johnson, A. B. Mindler, and R. H. Schlossel. "The Use of Ion Exchange for Treating Wastewaters and Recovering Metal Values from Electroplating Operations," *AIChE Symp. Ser. 82*(243, *Sep. Heavy Met. Other Trace Contam.*): 67–74 (1985).

I-59. Stortini, R. "Purification of Effluents of Acid Copper and Nickel Plating Galvanic Processes with Conventional Cation-Exchange Resins," *NATO ASI Ser.* Ser. E, 98 (Fundam. Appl. Ion Exch.): 34–40 (1985).

I-60. Yamasaki, K., T. Morikawa, and S. Eguchi. "Removal and Recovery of Copper from Waste Rinse Water of Chemical Coating by Ion Exchange," *Osaki-Furitsu Kogyo Gijutsu Kenkyusho Hokoku* (87):26–32 (1985).

I-61. Brooks, C. S. "Metal Recovery from Waste Acids," in *Proceedings of the 40th Purdue Industrial Waste Conference* (Boston: Butterworth, 1985), pp. 551–559.

I-62. Cao, Y. "Combined Ion-Exchange and Electrolysis Method for Treating Copper-Containing Effluent," *Diandu Yu Yuanbao* 5(3): 21–22 (1985).

I-63.   Hubicki, Z. "Copper Recovery from Ammonia Waste-waters on Various Types of Ion Exchangers," *Rudy Met. Niezelaz.* 30(9):338–343 (1985).

I-64.   Saieva, C. J. "Process and Apparatus for Recovering Metals from Dilute Solutions." U.S. Patent 4652352 (1987).

I-65.   Skvortsov, N. G., V. Y. Vasin, and G. M. Kolosova. "Nickel Recovery from Wash Waters of Nickel Electroplating Baths on Cation Exchangers KB-4," *Vestn. Mashinostr.* (4):56–57 (1987).

I-66.   Wang, F. H. and T. J. Hsu. "Removal and Recovery of Copper from Plating Wastewater with Cation-Exchange Resins," *Huan Ching Pao Hu (Taipei)* 10(1):35–46 (1987).

I-67.   Brown, C. J. and M. J. Dejak. "Process for Removal of Copper from Solutions of Chelating Agent and Copper." U.S. Patent 4666683 (1987).

I-68.   Diaz, M. and F. Mijangos. "Metal Recovery from Hydrometallurgical Wastes," *J. Metals* 39(7):42–44 (1987).

I-69.   Ichimura, K., T. Hirotsu, M. Sakuragi, N. Morii, and M. Suda. "Chelating Resins for Metal Ion Recovery." *Jpn. Kokai Tokkyo Koho,* Japanese Patent 62048725 (1987).

I-70.   Jin, T. "Treatment of Electroplating Wastewater by Air Flotation and Ion Exchange Method," *Diandu Yu Huanbao* 7(4):36–38 (1987).

I-71.   Dejak, M. and T. Nadeau. "Copper, Nickel, and Chromium Recovery in a Jobshop to Eliminate Waste Treatment and Sludge Disposal," *Hazard. Waste Hazard. Mater.* 4(3):261–271 (1987).

I-72.   Ruml, V. and M. Soukup. "Recovery of Copper, Nickel, Zinc, and Cadmium from Rinse Water of Electroplating Process." Czechoslovakian Patent CS 238173 (1987).

I-73.   Kauspediene, D. and H. Laumenskas. "Sorption of the Nonionic Surfactant Syntamid-5 on Ion-Exchange

Resins," *Liet. TSR Mokslu Akad. Darb. Ser. B* (3):41–47 (1987).

I-74. Filipov, P. and V. Bakushev. "Utilization of Valuable Components from Catalyst Production Wastewaters," *Metalurgiya (Sofia)* 42(40):24–26 (1987).

I-75. Rustamov, S. M. and F. T. Makhmudov. "Concentration of Silver and Nickel Ions from Wastes on Sodium-Clinoptilolite," *Zh. Prikl. Khim. (Leningrad)* 61(1):34–37 (1988).

I-76. Dejak, M. "Ion Exchange + Electrowinning = Recovery at Hewlett Packard," *Plating Surf. Finish.* 75(4):35–38 (1988).

I-77. Schlossel, R. H. "Metal-Containing Wastewater Treatment and Metal Recovery Process." U.S. Patent 4756833 (1988).

I-78. Jeffers, T. H., K. S. Gritton, P. G. Bennett, and D. C. Seidel. "Recovery of Cobalt from Spent Copper Leach Solution Using Continuous Ion Exchange," *Nucl. Chem. Waste Manage.* 8(1):37–44 (1988).

I-79. Brown, C.J. "A Better Way to Recover Nickel," *Prod. Finish. (Cincinnati)* 52(11):54–64 (1988).

I-80. Kalalova, E., Z. Kovarova, and V. Brozek. "Recovery and Separation of Cobalt, Nickel, Manganese, and Iron After Acid Leaching." Czechoslovakian Patent CS 249560 (1988).

I-81. Huang, C. Y. and S. C. Chen. "Recovery of Copper from Electroplating Waste Liquors by Ion Exchange on Casein," *Huan Ching Pao Hu (Taipei)* 11(2):10–12 (1988).

I-82. Zhang, Y. and K. Hang. "Recovery Technology and a Comprehensive Wastewater Treatment in a Nickel-Copper Smelter," *Huanjing Baohu (Beijing)* 9:15–16 (1988).

I-83. Schlossel, R. H. "Metal-Containing Wastewater Treatment and Metal Recovery Process." U.S. Patent 87–64520 (1989).

I-84. Celi, A. "Recovery of Metals from Wastes Containing Plastics." German Patent DE 3732177 (1989).

I-85.   Li, S., F. Wu, L. Zhang, and K. Xiao. "Study on Recovery of Nickel in the Waste Liquor of Nickel Coating," *Huanjing Baohu (Beijing)* (7):2–4 (1988).

I-86.   Kieszkowski, M., P. Ciecko, and R. Waskiak. "Recovery of Copper from Effluents and/or Waste Solutions Containing Stable Complexes from Printed-Circuit Manufacture." Polish Patent PL 144257 (1988).

I-87.   Suwa, T., N. Kuribayashi, and T. Yasumune. "Cerium Recovery from Waste Solutions Formed in Radioactivity Decontamination." *Jpn. Kokai Tokkyo Koho*, Japanese Patent JP 01101500 (1989).

I-88.   Sricharoenchaikit, P. "Ion Exchange Treatment for Electroless Copper-EDTA Rinse Water," *Plating Surf. Finish.* 76(12):68–70 (1989).

I-89.   Jenneret-Gris, G. "Chelating Resins and Their Use in the Extraction of Metals." Swiss Patent WO 8909238 (1989).

I-90.   Holl, W., S. Eberle, and J. Horst. "Process for Removing Heavy Metal Cations and/or Alkali Metal Cations from Aqueous Solutions with an Ion Exchanger Material." U.S. Patent 4894167 (1990).

I-91.   Vignola, M. A. "Removal of Metal Complexes from Wastewaters by Anion Exchange." German Patent DE 3614061 (1986).

I-92.   Uhlemann, E., W. Mickler, and H. Burghardt. "Recovery of Copper from Electrolysis Wastewater and Concentrate." East German Patent DD 256528 (1988).

I-93.   Alexandratos, J. D. "Synthesis of Dual Mechanism Ion Exchange/Redox Resins and Ion Exchange/Coordinating Resins with Application to Metal Ion Separations," in *Metals Speciation Separation and Recovery*, J. W. Patterson and R. Passino, Eds. (Chelsea, MI: Lewis Publishing Co., 1987), p. 527.

I-94.   Grole, M., M. Sandrock, and A. Kettrup. "The Separation of Precious Metals by Polymers Functionalized with Dehydrodithizone Derivatives," in *Metals Speciation Separation and Recovery*, J. W. Patterson and R.

Passino, Eds. (Chelsea, MI: Lewis Publishing Co., 1987), p. 551.

I-95.   Etzel, J. E. and D. H. Tseng. "Cation Exchange Removal of Heavy Metals with a Recoverable Chelent Regenerant," in *Metals Speciation Separation and Recovery*, J. W. Patterson and R. Passino, Eds. (Chelsea, MI: Lewis Publishing Co., 1987), p. 571.

I-96.   Kennedy, D. C., A. P. Becker III, and A. A. Worcester. "Development of an Ion Exchange Process to Recover Cobalt and Nickel from Primary Lead Smelter Residues," in *Metals Speciation Separation and Recovery*, J. W. Patterson and R. Passino, Eds. (Chelsea, MI: Lewis Publishing Co., 1987), p. 593.

I-97.   Gold, H., G. Czupryna, R. D. Levy, C. Coleman, and R. L. Gross. "Purifying Plating Baths by Chelate Ion Exchange," in *Metals Speciation Separation and Recovery*, J. W. Patterson and R. Passino, Eds. (Chelsea, MI: Lewis Publishing Co., 1987), p. 619.

I-98.   Diaz, M. and F. Mijangos. "Metal Recovery from Hydro-metallurgical Wastes," *J. Metals* 39(7):42 (1987).

I-99.   Karelin, Y. A. and E. P. Yakubovskii. "Treatment of Wastewater from Alkaline Etching of Printing Plates," *Vodosnabzh. Sanit. Tekh.* (11):5–7 (1985).

I-100. Kim, B. M. "Method and Apparatus for Continuous Ion Exchange." U.S. Patent 45623337 (1986).

I-101. Manczak, M. "Recovery of Copper from Etching Wastewater," *Gaz, Woda Tech. Sanit.* 54(5):140–142 (1980).

I-102. Annusewicz, A. "Electrolytic Recovery of Nickel from Washing Wastewaters," *Powloki Ochr.* 9(4–5):36–40 (1981).

I-103. Petkova, E., K. Vasliev, D. Petrov, V. Shkodrova, A. Tsanev, and K. Atanasov. "Extraction of Sulfuric Acid and Nickel from a Waste Copper Electrolyte," *Metalurgiya (Sofia)* 36(6):11–13 (1981).

I-104. Jellinek, H. H. G. and S. P. Sangal. "Complexation of Metal Ions with Natural Electrolytes," *Water Res.* 6(3):305–314 (1972).

I-105. Fuest, R. W. and M. J. Smith. "Feasibility Study of Regenerative Fibers for Water Pollution Control." U.S. EPA Water Pollution Control Research Series #17040 DFC (1970).

I-106. Pope-Reid Associates, Inc. "Central Recovery Facility Feasibility Study." Prepared for Minnesota Department of Energy Planning & Development, Technical Assistance Report, U.S. Department of Commerce, November 30 (1982).

I-107. Gupta, A. et al. "A Central Metal Recovery Facility." Prepared for New Jersey Department of Environmental Protection, Division of Water Resources (October 1983).

5.5 MEMBRANE SEPARATIONS

There are several types of membranes used for metal separations: electrodialysis, reverse osmosis, donnan membrane, ultrafiltration, and several kinds of special liquid membranes. In any electrodialysis cell there are ionic membranes adjacent to the electrodes, anionically charged (sulfonate exchange resin) for the anode and cationically charged (quaternary ammonium resin) for the cathode. In reverse osmosis application of pressure moves solvent water through a semipermeable membrane to concentrate charged metal solutes species. Donnan membranes achieve separations based on differences in the diffusion rates for the solution components. Ultrafiltration achieves separation based on the filtering effects of the membrane pores relative to the molecular size of solute species under imposed hydrostatic pressure. One type of special liquid membrane consists of interfacial surfactant films in an emulsion.

Among the 56 literature citations, membrane separations have been applied to 10 metals, namely: Ag, Al, Cd, Co, Cr,

Table 5.6 Metal Separation with Membranes

Waste system	Metals	Membrane	Separation process	Ref.
Electronic	Cu, Ag		Electrodialysis	MS–15
Plating	Ni		Reverse osmosis	MS–16
Bumper	Ni		Diffusion dialysis	MS–19
Metal waste	Ni	Membrane LIX 64N	Solvent extraction LIX agent in membrane	MS–18
Metal waste	Cu	NiFerrite anode (Selemion AMV anionic membrane; Selemion CMV cationic membrane)	Electrodialysis	MS–20
Metal waste	Cu	Polyethylene	Electrodialysis	MS–22
Ni	Ni	Hollow tube	Donnan–dialysis	MS–23
Plating	Ni	Ion exchange membrane	Electrodialysis	MS–27
Plating	Cr	Cation selective membrane	Electrodialysis	MS–28
Plating	Cd, Cr, Cu, Zn	Polyethylenimine membrane	Ultrafiltration	MS–32
Wastewater	Ni	Polyethylenimine	Electrodialysis	MS–33
Wastewater	Cu	Porous polymer membrane	Cu complexes with LIX65N Acorga P5300, P5100, 2EHPA	MS–37
Plating waste	Cr, Ni, Cd	Not specified	Electrodialysis	MS–39
Wastewater	Cu	PTFE membrane	Membrane osmosis	MS–40
Acid wastewater	Ag, Al, Co, Cd, Cu	PTFE membrane	Solvent extraction	MS–41

Source	Metal	Membrane/Material	Process	Code
Metal waste	Ni	Anion-selective membrane (MA3475)	Electrodialysis	MS-42
Ni waste	Ni	Cation-exchange membrane	Electrodialysis	MS-43
Acid wastewater	Cu	Dialysis membrane	Electrodialysis 90-95% Cu separation	MS-44
Plating	Cu	Acrylic-alkylsulfonic acid copolymer	Ultrafiltration	MS-45
Plating	Pb, Cu, Ni, Zn	Dialysis membrane	Electrodialysis	MS-46
Plating waste	Various metals	Not specified	Reverse osmosis	MS-47
Cu waste	Cu	Liquid membrane (LIX84 + decane)	Liquid membrane extraction	MS-48
Cu waste	Cu	Nafion 427 ion-exchange membrane	Electrodialysis	MS-49
Wastewater	Cu, Ni, Cr	Anion/cation exchange membrane	Ultrafiltration	MS-50
Industrial waste	Co, Mn, Ni, Cr	Liquid emulsion membranes (TBP + SPAN)	Solvent extraction	MS-51
Metal solutions	Cu and various metals	Anionic (Na dodecyl sulfonate) micelles	Ultrafiltration	MS-55

Cu, Hg, Mn, Ni, and Zn. Electrodialysis is the most commonly used separation process, with 21 citations. Reverse osmosis is the second most commonly used metal separation process, with 8 citations. Five examples of ultrafiltration were used for metal separations. There were only three examples of metal separation by uncharged donnan membranes. Selected results for the separation processes applied to various waste systems are summarized in Table 5.6.

The membrane separation processes are most appropriate for dilute solutions such as rinse waters. Most membranes are fragile mechanically and vulnerable to degradation by corrosive and strongly oxidative systems such as concentrated chromic acid. Separation efficiency also requires elimination of insolubles and suspended solids which block the membrane surfaces. There is much current work in progress to overcome these limitations and take advantage of the many advantageous separation potentials of the membrane systems.

References

MS-1.   Michaels, A. S. "New Separation Technique for the CPI," *Chem. Eng. Progress* 64:31–43 (1968).
MS-2.   Lacey, R. E. "Membrane Separation Processes," *Chem. Eng.* (Sept. 4, 1972), pp. 56–74.
MS-3.   Channabasappa, K. C. "Use of Reverse Osmosis for Valuable By-Products Recovery," in *Chemical Engineering Progress Symposium Series 107*, Vol. 67: 250–259 (1970).
MS-4.   Lonsdale, H. K. and H. E. Podall, Eds. *Reverse Osmosis Membrane Research* (New York: Plenum Press, 1972).
MS-5.   Golomb, A. "Application of Reverse Osmosis to Electroplating Waste Treatment," *Plating* 57:376 (1970).
MS-6.   Golomb, A. "Application of Reverse Osmosis to Electroplating Waste Treatment. I. Recovery of Nickel," *Plating* 57:1001 (1970).

MS-7. Golomb, A. "Application of Reverse Osmosis to Electroplating Waste Treatment. III. Pilot Plant Study and Economic Evaluation of Nickel Recovery," *Plating* 60:482 (1973).

MS-8. Spatz, D. D. "A Case History of Reverse Osmosis Used for Nickel Recovery in Bumper Recycling," *Plating Surf. Finish.* (July 1979).

MS-9. Lakshminarayanaiah, N. *Transport Phenomena in Membranes* (New York: Academic Press, 1969).

MS-10. Kesting, R. E. *Synthetic Polymeric Membranes.* (New York: McGraw-Hill Book Company, 1971).

MS-11. Perry, E. S. and C. J. Van Oss. *Progress in Separation,* Vol. 3 (New York: Wiley-Interscience, 1970).

MS-12. Danesi, P. R., E. P. Horowitz, G. F. Vandegriff-Argonne, and R. Chiarizid. "Mass Transfer Rate Through Liquid Membranes: Interfacial Chemical Reactions and Diffusion as Simultaneous Permeability Controlling Factors," *Sep. Sci. Technol.* 16(2):1201 (1981).

MS-13. Izaff, R. M., D. V. Dearden, D. W. McBrode, Jr., J. L. Oscarson, J. D. Lamb, and J. J. Christensen. "Metal Separations Using Emulsion Liquid Membranes," *Sci. Technol.* 18(12–13):1113 (1983).

MS-14. Danesi, P. R. "Separation of Metal Species by Supported Liquid Membranes," *Sep. Sci. Technol.* 19(11–12):857 (1984–85).

MS-15. Kubota, N., M. Nagata, and Y. Uchino. "Investigation on the Treatment of Effluents from Electronic Component Parts Plating Plants and the Recovery of Valuables. IV. Concentrating rinse waters by electrodialysis," *Kumamoto-ken Kogo Shikenjo Kenkyu Hokoku,* 52:18–23 (1977).

MS-16. Peking Institute of Environmental Protection; City of Peking. "Reverse Osmosis Technique and Its Application in the Treatment of Nickel Plating Waste Water," *Huan Ching K'o Hsueh* (3):65 (1978).

MS-17. Ueshima, H. "Recovery of Heavy Metals in Waste Water," *PPM* 10(3):18–31 (1979).

MS-18. Baker, R. W. "Removal of Metal Ions from Aqueous Solutions with Separation Membranes." Japanese Patent 54017381 (1979).

MS-19. Utsunomiya, T. and H. Shibata. "Recovering Acids and Metals from Waste Solutions." Japanese Patent 51040303 (1976).

MS-20. Araki, S., I. Yokoyama, H. Komatsu, and A. Koyama. "Recovery of Metallic Copper from Acidic Waste Solution." Japanese Patent 54104439 (1979).

MS-21. Chen, C.-C., D.-H. Lin, and Y.-M. Chen. "Study on the Treatment of Wastewater from Effluent Streams. I. Membrane Electrodialysis of Nickel Solution," *Hua Hsueh* (2):31–36 (1979).

MS-22. Catonne, J. C. "Application of Ion-Exchange Membranes for Recycling of Rinse Water After Alkaline Copper Plating," *Galvano-Organo* 50(516):489–490 (1981).

MS-23. Ng, P. K. and D. D. Snyder. "Mass Transport Characterization of Donnan Dialysis: the Nickel Sulfate System," *J. Electrochem. Soc.* 128(8):1714–1719 (1981).

MS-24. Eisenmann, J. L. "Method and Apparatus for Recovering Charged Ions from Solution." European Patent 34661 (1981).

MS-25. Eisenmann, J. L. "Nickel Recovery from Electroplating Rinsewaters by Electrodialysis." EPA-600/2–81–130, No. PB81–227209 (1981).

MS-26. Markovac, V. and H. C. Heller. "Engineering Aspects of Electrodialysis for Nickel Plating Rinsewater," *Plating Surf. Finish.* 69(1):84–87 (1982).

MS-27. Huang, T.-C. and J.-Y. Chou. "Recovery of Metal Ion from Waste Electroplating Solution," *K'o Hsueh Fa Chan Yueh K'an*, 9(12):1069–1079 (1981).

MS-28. Altmayer, F. "Introducing the COPS. One-Step Regeneration and Purification of Chromic Pickling/Stripping Solutions," *Plating Surf. Finish.* 70(3):20–24 (1983).

MS-29. Heller, H. C. and Markovac, V. "Identification of a Membrane Foulant in the Electrodialytic Recovery of Nickel," *Anal. Chem.* 55(4):551A-551A, 554A, 556A-557A (1983).

MS-30. Sumitomo Metal Mining Co. Ltd. "Recovery of Nickel Sulfate from Copper Electrowinning Waste Solutions." Japanese Patent 59056590 (1984).

MS-31. Sharipov, M. S. and A. A. Zharmenov. "Use of Membrane Process in the Waste-Free Manufacture of Copper," *Vestn. Akad. Nauk Kaz. SSR* (5):43–46 (1984).

MS-32. Dytnerskii, Y. I., Y. N. Zhilin, K. A. Volcheck, and V. S. Pshezhetskii. "Extraction of Metals from Natural and Wastewaters by Complexing and Ultrafiltration," *Khim. Promst. (Moscow)* (8):477–479 (1984).

MS-33. Deuschle, A. and E. Kuebler. "Electrodialysis – Recovery of Materials from Electroplating Rinse Waters," *Galvanotechnik* 78(5):968–91 (1984).

MS-34. Draxler, J. and R. Marr. "Recovery of Metal Ions," *Chem.-Anlagen Verfahren* 17(Sonderausg. E3: Edition Europa):45 (1984).

MS-35. Binder, H. and H. Behret. "Construction Principles of Electrodialysis Apparatus, and Recovery of Raw Materials by Electrodialysis," *DECHEMA-Monogr.* 97(*Elektrochem. Verfahrenstech.*):289–303 (1984).

MS-36. Asahi Chemical Industry Co. Ltd. "Separation of Metal Ions from Aqueous Solutions." Japanese Patent 60061005 (1985).

MS-37. Schuegeri, K., A. Mohrmann, W. Gutknecht, and H. B. Hauertmann. "Application of Liquid Membrane Emulsion for Recovery of Metals from Mining Wastewaters and Zinc Liquors," *Desalination* 53:197–215 (1985).

MS-38. Roach, E. T. "Evaluation of Donnan Dialysis for Treatment of Nickel Plating Rinsewater." EPA/600/2-85/055, No. PB85-200046 (1985).

MS-39. Pekin, T. and R. J. Capabo, Jr. "Pretreatment and Recovery at a Plating Shop," *Annual Technical Confer-*

ence Proceedings—American Electroplating Society 72nd (Sess. E), Paper E2 (1985).

MS-40. Eto, Y. and I. Kato. "Recovery of Metals from Aqueous Solutions." Japanese Patent 61143527 (1986).

MS-41. Eto, Y. and Y. Taniguchi. "Recovery of Metal in Acidic Wastewater by Extraction." Japanese Patent 61132516 (1986).

MS-42. Verbaan, B. "Neutralization of Acid Contained in Aqueous Acid Solutions." South African Patent 8509471 (1986).

MS-43. Higashi, K. and S. Nagashima. "Recovery of Nickel from Plating Rinse Water by Electrolysis with Diaphragm," *Kenkyu Hokoku—Tokyo-toritsu Kogyo Gijutsu Senta* (15):126–130 (1986).

MS-44. Abe, T., T. Yoshihara, T. Sano, and T. Mimuro. "Apparatus for Treatment of Acid-Pickling Wastewater and Residue." Japanese Patent 62020838 (1986).

MS-45. Wellner, P., V. Haenel, W. Maier, H. Waeschke, G. Malsch, and H. Roedicker. "Recovery of Copper from Acid Using Baths." German Democratic Republic Patent 249287 (1987).

MS-46. Tholen, J. P. P. and E. J. Rijkoff. "Membrane Electrolysis of Electroplating Wastewater," *PT-Procestech.* 42(12):123 (1987).

MS-47. Carwright, P. S. "Application of Reverse Osmosis to Metal Recovery," *Galvano-Organo-Trait. Surf.* 57(587):541–543 (1988).

MS-48. De Haan, A. B., P. V. Bartels, and J. De Grauw. "Extraction of Metal Ions from Wastewater Streams," *I²-Procestechnologie* 4(9):20–23, 25–26 (1988).

MS-49. Meyyappan, R. M., N. Sathaiyan, and P. Adaikkalam. "Recovery of Copper from Ammoniacal Copper Etchants Using Ion-Exchange Membranes," *Bull. Electrochem.* 5(2):121–123 (1989).

MS-50. Vigo, F., C. Uliana, and M. Traverso. "Use of Ultra-filtration for Effluents Containing Inorganic Ions," *ICP* 17(6):106–111 (1989).

MS-51. Abou-Nemeh, I. and A. P. Van Peteghem. "Extraction of Cobalt and Manganese from an Industrial Effluent by Liquid Emulsion Membranes," in *Proceedings of the International Conference Separation Science and Techynology, 2nd,* M. H. Baird, and S. Vijayan, Eds. (Ottawa, Canada: Canadian Society for Chemical Engineering, 1989), 416–423.

MS-52. Hockanser, A. M. "Concentrating Chromium with Liquid Surfactants," *AIChE Symp. Ser.* 71(152):136 (1975).

MS-53. Warnke, J. E., K. G. Thomas, and S. C. Creason. "Reclaiming Plating Waste Water by Reverse Osmosis," in *Proceedings of the 31st Purdue Industrial Waste Conference* (Ann Arbor, MI: Ann Arbor Science, 1976).

MS-54. Strzelbicki, J. and W. Charewica. "Separation of Copper by Liquid Surfactant Membranes," *J. Inorg. Nucl. Chem.* 40 (1978).

MS-55. Scamehorn, J. F., S. D. Christian, and R. T. Ellington. "Use of Micellar-Enhanced Ultrafiltration to Remove Multivalent Metal Ions from Aqueous Streams," in *Surfactant-Based Separation Processes,* J. F. Scamehorn and H. H. Harwell, Eds. (New York: Marcel Dekker, Inc., 1989) Chapter 2.

MS-56. O'Hara, P. A. and M. P. Bohrer. "Application of a Liquid Membrane for Copper Recycling," in *Chemical Separations,* Vol. II, C. J. King, Ed. (Denver: Litarrian Literature, 1986).

5.6 SOLVENT EXTRACTION

The recovery of metals in solution by solvent extraction has been used to achieve analytical separations, a wide range of metallurgical separations, and more recently in the treatment of wastewaters to remove soluble metals.[S-1 to S-70] A considerable body of technology is applicable to separation of mixtures of metals, such as cobalt, cadmium, copper, nickel, molybdenum, chromium, vanadium, zinc, uranium, etc., from aqueous solution into an immiscible organic phase by formation of organic salts or chelate compounds which provide a favorable solubility distribution between the aqueous and organic phases.

Some of the general types of compounds applicable to solvent extraction consist of dicarboxylic acids, aliphatic amines, aromatic amines, amino acids, hydroxy acids, alkyl phosphates, nitroacids, salicylaldehyde derivatives, hydroxy aldehydes, phenolic compounds, and O,O'-dihydroxy azo dyes, to mention only a few of the reactants. Comprehensive discussions of reaction mechanisms, metal-organic reactant combinations, and extraction conditions are covered by Bailar,[S-1] Morrison and Freiser,[S-2] Stary,[S-3] Katzin and Lagowski,[S-4] and Marcus and Kertes.[S-5] The term "liquid ion exchange" is used commonly when the reaction mechanism occurs by a cation-anion combination.[S-6] There is an extensive literature on solvent extraction separations and concentrations applied to extractive metallurgy.[S-6-S-25] Additional relevant studies have used solvent extraction for metal removal from wastewaters, but in most cases without metal recovery as the objective.[S-26-S-31]

Nonselective removal of contaminant metals from mixtures of metals in aqueous solution can be achieved by solvent extraction with a variety of organic reactants. Some examples of such agents are 8-hydroxyquinoline for cobalt, molybdenum, copper, iron, nickel, cadmium, zinc, vanadium, manganese, lead, and titanium. In some instances highly specific extractants can be selected, such as dimethyl glyoxime for nickel, can be used for metal recovery. In these applications, only palladium seriously interferes. Solvent extraction can be advantageously combined with electrodeposition, ion exchange, or precipitation to provide selective separation and concentration of individual metals.

Successful application of available solvent extraction technology requires development of technically feasible, economical processes for recovery of metals such as cadmium, cobalt, copper, chromium, nickel, lead, tin, and zinc from mixtures with metals such as iron or aluminum as undesirable contaminants. Summaries of the types of chemical agents available and applicability to the metals of interest are shown in Table 5.7. Figures 5.1 and 5.2 show the pH dependence for agents of demonstrated performance for the important metals copper and nickel. Selectivity in separations is quite dependent on pH and maximum advantage must be taken of this parameter in combination with the type of extraction agent chosen. There are some examples of uniquely high selectivity for metals such as nickel and palladium for dimethyl glyoxime as the extraction agent.

There are demonstrations at bench scale that sequential solvent extractions at pH less than 3 using DI2EHPA for iron and oxime agents such as LIX63, LIX64, or LIX622 for copper permit efficient separation of copper from metal mixtures.[S-51] Efficient selective separations for other metals such as Cd, Co, Cr(III), Ni, and Zn are more difficult to achieve, but there are promising agents that can be applied for making selective sequential solvent extractions of these latter metals, possibly in combination with alternative processes such as selective precipitation and/or ion exchange (References S-18 to S-20, S-23, S-24, S-37, S-41, S-43, S-45 to S-47, S-52, S-54, S-56, S-57, S-61 to S-66, S-68).

Table 5.7 Metal Recovery by Solvent Extraction

Waste	Metals	Extraction agent	Recovery efficiency %	Ref.
Copper ore	Co, Ni, Cu	Linoleic, oleic, erucic, palmitic acid		S-35
Metal solution	Cu, Zn	SME529, α-hydroxyoxime (Cu) Bis(2-ethyl)phosphoric acid (Zn)		S-37
Metal solution	Cu, UO_2	C16–18 fatty acids carrier Bis(2-ethylhexylphosphate) Trioctyl phosphine oxide		S-38
Mine water	Cu, Zn, Fe	Versatic 10 (carboxylic acid)	99.9 Cu/Fe 99 Zn	S-40
Spent catalysts	Ni, Co, Mo, W, Cu, Cr	Kelex 100 + LIX 64	73-99 Cr, Cu, Ni, Mo, W	S-47
Metal waste	Cu, Cr, Cd, Al, Ni, Pb, Sn, Zn, Fe	Di-2-ethylhexylphosphoric acid and LIX 63		S-51
Petroleum coke	Ni, V	20 D2EHPA + TBP	85 V 73 Ni	S-52
Boiler slag	V		94 V	S-53
Industrial waste	Ni	D2EHPA		S-54
Hydromet. waste	Co, Cr, Mn, Ni, W	Amberlite LA-2 Triisooctylamine, decanal oxime	97 Co, 79 Cr, 99 Mn, 99 Ni, 90 W	S-56

Source	Metals	Reagent	Purity	Ref.
Metal waste	Ni	Bis-2-ethylhexyldithiophosphoric acid	99.85 Ni	S-57
Nonferrous waste	Cu, Cr, Al, Ni, Zn	DI3EHPA, hydroxyoximes, isononanoic acid		S-58
Battery waste	Li	LIX 51 (α-perfluro-alkanoyl-m-dodecylacetophenone + trioctyphosphine oxide)	75-99 Li	S-59
Wastewater	Cu, Co, Ni	3,4-Dimethylyaniline, aniline, O-methylaniline, O-chloroaniline, p-methylaniline, 2,4 dimethylaniline		S-61
Battery waste	Cd, Ni	MSP-8-(alkylmonothiophosphonic acid)	100 Ni 89 Cd	S-62
Spent catalyst	Co, Ni, Mo, W	Caustic extraction after chlorination		S-63
Metal waste	Ce, Ge, Co, Mn, U, Cu, Pd, Ni	Amidoxime, imidodioxime		S-64
Plating waste	Co, Cr, Ni	D2EHPA		S-65
Plasting waste	Fe, Cu, Ni, Cr, Zn	Cyclohexane nonoxime		S-66
Metal wastes	Co, Mn, Ni	2-Hydroxy-3-chloro-5-nonyl benzophenone + decanoic acid	98 Co	S-43
Ores leach liquors	Co, Ni	Phosphonic or phosphinic aids (Shell SME 414, American Cyanimid Cyanex 272)		S-68

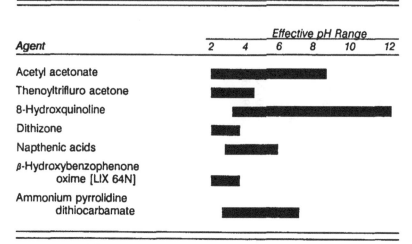

Figure 5.1.   Solvent extraction agents for copper—pH dependence.

Agent	Effective pH Range					
	2	4	6	8	10	12
8-Hydroxyquinoline						
Diethyldithiocarbamate						
Naphthenic acids						
Dimethylglyoxime						
α-Nitrosonaphthol						
Ammonium pyrrolidine dithiocarbamate						

Figure 5.2.   Solvent extraction agents for nickel—pH dependence.

References

S-1.   Bailar, J. C., Jr. *The Chemistry of the Coordination Compounds.* ACS Monograph (New York: Reinhold, 1956).

S-2.   Morrison, G. H. and H. Freiser. *Solvent Extraction in Analytical Chemistry* (New York: John Wiley & Sons, Inc., 1957).

S-3. Stary, J. *The Solvent Extraction of Metal Chelates* (New York: Pergamon Press, Macmillan, 1964).

S-4. Katzin, L. I. and J. J. Lagowski, Eds. *The Chemistry of Nonaqueous Solvents* (New York: Academic Press, Inc., 1966), pp. 173–205.

S-5. Marcus, Y. and A. S. Kertes. *Ion Exchange and Solvent Extraction of Metal Complexes.* New York: Wiley Interscience (1969).

S-6. Li, N. N., Ed. "Liquid Ion Exchange in Hydro Metallurgy," in *Recent Developments in Separation Science*, Vol. 2 (Cleveland: CRC Press, Inc., 1975).

S-7. Baes, C. F. and H. T. Baker. "Extraction of Fe(III) from acid perchlorate solutions by Di(ethyl-hexyl) phosphoric acid in n-octane," *J. Phys. Chem.* 64:89 (1960).

S-8. Fletcher, A. M. et al. "Solvent extraction of ferric iron by a carboxylic acid," *Bull. Inst. Min. Met.* 696:81–88 (1964).

S-9. Fletcher, A. W. and D. S. Flett. "Equilibrium Studies on the Solvent Extraction of Some Transition Metals with Naphthenic Acid," *J. Appl. Chem.* 14:250 (1964).

S-10. Fletcher, A. W. and K. D. Hester. "A New Approach to Copper-Nickel Ore Processing," *Trans. AIME* 229:282–291 (1964).

S-11. Fletcher, A. W. and J. C. Wilson. "Naphthenic Acid as a Liquid-Liquid Extraction Reagent for Metals," *Bull. Inst. Min. Met.* 652:355 (1961).

S-12. Brooks, P. T. and J. B. Rosenbaum. "Separation and Recovery of Cobalt and Nickel by Solvent Extraction and Electrorefining." U.S. Bureau of Mines RI 6159 (1963).

S-13. Bridges, D. W. and J. B. Rosenbaum. "Metallurgical Applications of Solvent Extraction. I. Fundamentals of the Process." U.S. Bureau of Mines Circ. 8139 (1962).

S-14. Rosenbaum, J. B., et al. "Metallurgical Applications of Solvent Extraction. II. Practice and Trends." U.S. Bureau of Mines Circ. 8502, NTIS PB 198134 (1971).

S-15. Brooks, P. T. "Preparation of High Purity Nickel and Cobalt." U.S. Patent 3446720 (1969).

S-16. Brooks, P. T., et al. "Processing of Superalloy Scrap," *J. Metals* (November 1970), pp. 25–29.

S-17. McGarr, H. J. "Solvent Extraction Stars in Making Ultrapure Copper," *Chem. Eng.* 77:82 (1970).

S-18. Sandberg, R. G. and T. L. Hebble. "Cobalt and Nickel Removal from Zinc Sulfate Electrolyte by Solvent Extraction and Precipitation Techniques." U.S. Bureau of Mines RI 8320 (1978).

S-19. Nilsen, D. N. et al. "Solvent Extraction of Nickel and Copper from Laterite-Ammoniacal Leach Liquors." U.S. Bureau of Mines RI 8605 (1982).

S-20. Nilsen, D. N. et al. "Solvent Extraction of Cobalt from Laterite-Ammoniacal Leach Liquors." U.S. Bureau of Mines RI 8419 (1980).

S-21. Osseo-Asare, K. and M. E. Keeney. "Sulfonic Acids: Catalysis for the Liquid-Liquid Extraction of Metals," *Sep. Sci. Technol.* 15:999 (1980).

S-22. Lo, T. C. et al. *Handbook of Solvent Extraction* (New York: John Wiley & Sons, Inc. 1983).

S-23. Preston, J. S. and C. A. Fleming. "Recovery of Nickel by Solvent Extraction from Acidic Sulfate Solutions," in *Proceedings of the 3rd International Symposium Hydrometallurgy Research Development and Plant Practice*, K. Osses-Asare and J. D. Miller, Eds. (Warrendale, PA: Metallurgical Society of the AIME, 1983).

S-24. Preston, J. S. "Solvent Extraction of Nickel and Copper by Mixtures of Carboxylic Acids and Non-Chelating Oximes," *Hydrometallurgy* 11:105 (1983).

S-25. Burgard, M. et al. U.S. Patent 4499057 (1985).

S-26. Seeton, F. A. "Solvent Extraction Recovers Vanadium from Waste Stream," *Chem. Eng.* 71: 112 (1964).

S-27. McDonald, C. W. "Removal of Toxic Metals from Metal Finishing Wastewater by Solvent Extraction." EPA 600/2-78-011; PB 280563 (1978).

S-28. McDonald, C. W. and R. S. Bajwa. "Removal of Toxic Metal Ions from Metal-Finishing Waste Water by Solvent Extraction," *Sep. Sci.* 12:435 (1977).

S-29. Baes, C. F. "The Extraction of Metallic Species by Dialkyl Phosphoric Acid," *J. Inorg. Nucl. Chem.* 24:707 (1962).

S-30. Clevenger, T. E. and J. T. Novak. "Recovery of Metals from Electroplating Wastes Using Liquid-Liquid Extraction. *J. Water Pollut. Control Fed.* 55:984–989 (1983).

S-31. Courduvelis, C. et al. "A New Treatment for Wastewater Containing Metal Complexes," *Plating Surf. Finish.* 70:70 (1983).

S-32. Helfferich, F. " 'Ligand exchange': a novel separation technique," *Nature* 189:1001–1002 (1961).

S-33. Kadoki, H., S. Tenma, F. Ito, and N. Kawakami. "Separation and Recovery of Valuable Metals in Waste Catalyst Containing Molybdenum." Japanese Patent 53115603 (1978).

S-34. Kadoki, H., S. Tenma, F. Ito, and N. Kawakami. "Separation and Recovery of Valuable Metals in Waste Catalyst Containing Molybdenum." Japanese Patent 53117602 (1978).

S-35. Kopacz, S., B. Jerowska, A. Lipinska, K. Trybuszewska, T. Franczak, and B. Mucha. "Preparation of Some Metal Compounds by Extraction of Wastes from Copper Ore Processing." Polish Patent 85945 (1976).

S-36. Keyworth, D. A. and J. R. Sudduth. "Recovery of Metals from Bimetallic Salt Complexes." U.S. Patent 4153452 (1979).

S-37. Barthel, G., H. Fischer, and U. Scheffler. "Separation and Recovery of Copper and Zinc from Sulfuric Acid Solutions of Metal Salts by Solvent Extraction," *Erzmetall* 32(4):176–184 (1979).

S-38. Fuller, E. J. "Extraction of Metals from Dilute Solutions." German Patent 2950567 (1980).

S-39. Breister, S. and J. R. Reiner. "Use of Sodium Salt of 2-Mercaptopyridine-N-Oxide to Separate Gold from Acidic Solutions." U.S. Patent 4368073 (1983).

S-40. Svendsen, H. F. and G. Thorsen. "Mine Water Treatment by Solvent Extraction with Carboxylic Acids,"

Oslo Symp. 1982: Ion Exch. Solvent Extr., Pap., IV/26-IV/43. J. Frost Urstad and G. Borgen, Eds. (London, U.K.: Society of Chemical Industry, 1982).

S-41. Padolina, C. D. and G. Thorsen. "Separation and Recovery of Zinc, Copper, and Iron from Mine Water," *Kimika* 1-21 (1982).

S-42. Daido Chemical Industry Co., Ltd. "Apparatus for metal extraction." Japanese Patent 58100639 (1983).

S-43. Cosmen Schortmann, P. and E. Diaz Nogueira. "Separation of Manganese, Cobalt and Nickel from Aqueous Solutions by Solvent Extraction." Spanish Patent 517148 (1983).

S-44. Kim, B. M. "A Membrane Extraction Process for Selective Recovery of Metals from Wastewater," in *AIChE Symp. Ser. 81(243, Sep. Heavy Met. Other Trace Contam.):*126-132 (1985).

S-45. Itagaki, K., K. Kanazawa, M. Hayashi, K. Matsuda, and A. Yazawa. "Fundamental Study on the Recovery of Cobalt and Nickel from Iron-Based Molten Alloys by Using Metal Solvents," in *Recycle Second Recovery Met., Proceedings of the International Symposium*, P. R. Taylor, H. Y. Sohn, and N. Jarrett, Eds. (Warrendale, PA: Metallurgical Society, 1985), pp. 489-502.

S-46. Jong, B. W. and R. E. Siemens. "Proposed Methods for Recovering Critical Metals from Spent Catalysts," in *Recycle Second Recovery Met., Proceedings of the International Symposium*, P. R. Taylor, H. Y. Sohn, and N. Jarrett, Eds. (Warrendale, PA: Metallurgical Society, 1985), pp. 477-488.

S-47. Nyman, B. G. and L. E. I. Hummelstedt. "Extraction Process for the Separation and Recovery of Metals from Aqueous Solutions." Belgian Patent 902810 (1985).

S-48. Bolt, A., M. Tels, and W. J. T. Van Gemert. "Recovery of Pure Metal Salts from Mixed Heavy Metals Hydroxides Sludges," in *Recycl. Int. [Int. Recycl. Congr.], 4th*, K. J. Thome-Kozmiensky, Ed. (Berlin: EF-Verlag Engr. Umwelttech, 1984), pp. 1025-1031.

S-49. Djugumovic, S., M. M. Kreevoy, and T. Skerlak. "Membranes for Metal Benefication or Water Purification; the Concentration Profile Within the Membrane," *Kem. Ind.* 35(8):435–439 (1986).

S-50. Schimmel, G., R. Gradi, and F. Kolkmann. "Processing of Heavy Metal-Containing Residues from Refining of Crude Phosphoric Acid." Federal Republic of Germany Patent 3522822 (1987).

S-51. Brooks, C. S. "Applications of Solvent Extraction in Treatment of Metal Finishing Wastes," *Met. Finish.* 85(3):55–59 (1987).

S-52. Sakuta, Y., Y. Tanabe, A. Takano, T. Taskashashi, T. Fujiwara, and T. Maruyama. "Survey of Rare Metal Resources and Their Recovery and Refining. II. Recovery of Vanadium and Nickel from Residues of Petroleum Coke Combustion," *Hokkaidoritsu Kogyo Shikenjo Hokoku* (285):37–43 (1986).

S-53. Tsuboi, I., M. Tamaki, J. Ingham, and E. Kunugita. "Recovery of Vanadium from Oil-Fired Boiler Slag by Direct Leaching and Subsequent Solvent Extraction," *J. Chem. Eng. Jpn.* 20(5):505–510 (1987).

S-54. Ganguli, P., R. Raghunand, and R. K. Lal. "Recovery of Nickel from Industrial Wastes: Optimized Process Based on Liquid-Liquid Extraction," *Indian J. Technol.* 25(12):656–663 (1987).

S-55. Ciernik, J., E. Spousta, M. Stastny, D. Ambros, L. Preisler, J. Sramek, and P. Rysanek. "Copper Recovery with Regeneration of Liquors After Oxidation Polycondensation of 2,6-Xylenol." Czechoslovakian Patent 250949 (1988).

S-56. Redden, L. D., R. D. Groves, and D. C. Seidel. "Hydrometallurgical Recovery of Critical Metals from Hardface Alloy Grinding Waste: A Laboratory Study." U.S. Bureau of Mines Rep. Invest., RI 9210 (1988).

S-57. Marr, R., H. Lackner, H. J. Bart, and A. Nicki. "Separation of Metal and Especially Nickel Ions from Aqueous Waste Solutions." German Patent 8900444 (1989).

S-58. Rueckel, H. G., N. Amsoneit, F. Dietl, and T. Knoblauch. "Treatment of Wastes Containing Nonferrous Metals." Federal Republic of Germany Patent 3732242 (1989).

S-59. Kunugita, E., J. H. Kim, and I. Komassawa. "A Process for Recovery of Lithium from Spent Lithium Batteries," *Kagaku Kogaku Ronbunshu* 15(4):857–862 (1989).

S-60. Dobos, G., J. Laszio, and L. Homoki. "Processing of Nickel Battery Wastes." Hungarian Patent 47501 (1989).

S-61. Kalembkiewicz, J. and S. Kopacz. "Extraction Solvents for Separating Copper(II), Nickel(II) and Cobalt(II) from Wastewaters and Process Waters." Polish Patent 146619 (1989).

S-62. Tsuboi, I. and E. Kunugita. "Separation of Cadmium from Nickel with Alkylmonothiphosphonic Acid," in *Proceedings of the Symposium on Solvent Extr.* (Hamamatsu, Japan: Jpn. Assoc. Solvent Extr., 1988), pp. 187–192.

S-63. Jong, B. W., S. C. Rhoads, A. M. Stubbs, and T. R. Stoeling. "Recovery of Principal Metal Values from Waste Hydroprocessing Catalysts." U.S. Bureau of Mines Rep. Invest., RI 9252 (1989).

S-64. Matsuda, K., K. Ouchi, and I. Kosaka. "Chelating Agent for Recovery of Heavy Metals." Japanese Patent 01141816 (1989).

S-65. Magdics, A. and D. B. Stain. "Recovery of Metals from Electroplating Wastewaters by Extraction and Recycling of Metals and Extracts." European Patent 332447 (1989).

S-66. Tels, M. and J. P. Lotens. "Recovery of Pure Metal Salts from Mixed Metal Hydroxide Sludges," in *Proceedings of the National Waste Processing Conference, 9th* (Eindhaven, Netherlands, 1980), pp. 109–119.

S-67. Doyle-Garner, F. M. and A. D. Moneimius. "Hydrolic Stripping of Single and Mixed Metal-Versatic Solutions," *Met. Trans.* 163:671 (1985).

S-68. Jacobs, J. J., S. Behmo, M. Allard, and J. Moreau. *Nickel and Cobalt Extraction Using Organic Compounds. Applied Technology Series Vol. 6*, Pergamon Press, Elmsford, NY (1985).

S-69. Clevenger, T. E. and J. T. Novak. "Recovery of Metals from Electroplating Wastes Using Liquid-Liquid Extraction," *J. Water Pollut. Control Fed.* 55(7):984–989 (1983).

S-70. Petrovicky, J., P. Vejnar, and J. Haman. "Processing of Recycled Polymetallic Raw Materials by Ammoniacal Leaching," *Freiberg. Forschungsh. A,* A 746, 115–126 (1987).

5.7 PRECIPITATION

Precipitation was cited in 126 publications as a means of separating metals from aqueous solution, most commonly with the objective of removal from the aqueous phase rather than metal recovery. Waste systems cited included a wide variety, ranging from chemical and metallurgical process effluents, electrical, electronic, and plating wastes, scrap alloys, various leachates, waste acids, and alkalies, and spent catalysts. Twenty six metals (Au, Ag, Al, As, Ca, Cd, Co, Cr, Cu, Fe, Hg, Ir, Mg, Mn, Mo, Ni, Pb, Pd, Pt, Rh, Sb, Se, Sn, V, W, Zn), were cited, with copper and nickel appearing most frequently. Precipitation agents most commonly used include hydroxides, carbonates, sulfides, and phosphates used alone and in various combinations. In addition, organic precipitants and a number of special precipitants appear in the chemical literature of the past 10 years. A summary of the most frequently cited precipitants used with various metals is shown in Tables 5.8 and 5.9.

Soluble metals can be separated and concentrated as insoluble metal hydroxides and/or carbonates by precipitation with various alkaline reactants, such as lime, magnesia, $NaHCO_3$, Na_2CO_3, $(NH_4)_2CO_3$, NaOH, and NH_4OH. Most of the metals of interest, such as cobalt, copper, cadmium,

Table 5.8 Metal Separation by Precipitation

Waste	Metals	Precipitation agent	Recovery efficiency %	Ref.
Plating	Cd, Cu, Zn	Sulfide	>99 Cd, Cu, Zn	P-19
Metal finishing	Cu, Cr, Ni	NaOCl, NaOH, NaHSO$_3$	88 Cr, 88 Ni Zn	P-25
Wastewater	Cr, Ni	Na$_2$CO$_3$	98 Ni	P-27
Metal salt solution	Cu	CaCO$_3$	75–80 Cu	P-30
Electrolytic Cu sludge	Au, Ag, Cu, Se	Chlorination	99.7 Au	P-33
Cu/As compounds	As, Cu	Sulfide	99.9 As, 99 Cu	P-105
Cu/Al halides	Cu	Al	95 Cu	P-34
Mine waste	Al, Cu, Ca, Mg, Mn, Ni, Fe, Zn	Sulfide + hydroxide + oxidizing agent	>85 metals	P-37
Electroless Cu	Cu	NH$_3$	90–96 Cu	P-40
Printing	Cu	NH$_3$	99.5 Cu	P-43
V Waste solution	V	NaOH, KOH, Ca(OH)$_2$	90 V$_2$O$_5$	P-55
Electrical waste	Cu, Ni, W	Carbonate, hydroxide	98 Cu, 98 Ni, 100 W	P-59

Heavy oil	Ni, V	$NaClO_3$, NaOH, NH_4OH	60-95 Ni + V	P-63
Metal finishing	Cd, Cu, Cr, Ni, Zn	NaOH	93-98 Zn	P-68
Ferrite waste	Cu	Hydroxyl amine hydrochloride + NaOH	99.3 Cu	P-79
Industrial waste	Cu	$Na_2S_2O_3$	99.7 Cu	P-113
Wastewater	Ag	Chloride + Cu, Zn	92-96 Ag	P-92
Electrolysis waste	Co, Cu, Ni	H_2O_2, oxalic acid	93-99 Co, Cu, Ni	P-116

Table 5.9 Precipitation Separations (References)

Metal	Hydroxide	Carbonate	Hydroxide/carbonate	Sulfide	Staged precipitation
Au	P-51, P-106				
Ag	P-51, P-106				
Al	P-37, P-84		P-15		
As	P-111			P-49, P-105	
Cd	P-21, P-68, P-106				P-19, P-28
Co		P-107	P-64, P-94	P-47	
Cr	P-16, P-21, P-23, P-24, P-25, P-68, P-84, P-86		P-67		P-32
Cu	P-25, P-43, P-51, P-52, P-57, P-59, P-68, P-70, P-74, P-79, P-88, P-91, P-95, P-106	P-30, P-46, P-53, P-66, P-72, P-73, P-87	P-15, P-38, P-81	P-45, P-47, P-62, P-71, P-76, P-105	P-17, P-19, P-28, P-37
Fe	P-24, P-36, P-37, P-58, P-64, P-84, P-111	P-46	P-15		
Mg	P-37				
Mn	P-37	P-53	P-94	P-71	
Mo	P-52	P-53	P-67		

Ni	P-21, P-23, P-25, P-26, P-35, P-50, P-51, P-58, P-59, P-61, P-68	P-16, P-18, P-27, P-41, P-66, P-72, P-73, P-107	P-15, P-67, P-94	P-41	P-17
Pb	P-74, P-111, P-112	P-87, P-106		P-71, P-76	P-19, P-28
Sn	P-87	P-75			
Ti	P-84				
V	P-58, P-63				
W	P-59		P-67		
Zn	P-64, P-68, P-80, P-112	P-60, P-66, P-72	P-15, P-20, P-73	P-45, P-47, P-71	P-19, P-28, P-37

nickel, manganese, zinc, etc., precipitate at pH above 6 or 7, permitting some possibility for separation from ferric iron, which precipitates below pH 5. Due to the potential for coprecipitation and the ion exchange characteristics of the colloidal hydroxide precipitates, even distinct differences in pH for hydroxide formation are no guarantee of avoiding mixed precipitates when metal mixtures are in solution. The alkalies, such as Na_2CO_3, NaOH, $(NH_4)_2CO_3$ or NH_4OH, forming soluble neutralization by-products are preferable to the alkaline precipitants (lime or magnesia) if metal recovery is desired in order to avoid insoluble contaminate residues.

Soluble metals can be separated or concentrated also as sulfides using Na_2S, NaHS, H_2S, or FeS (Table 5.8) and have been applied frequently for metal removal from waste effluents to achieve emission standard limitations. Commercial sulfide precipitation processes are available for metal removal from aqueous waste effluents.[P-4-P-7] The lower solubility of the metal sulfides of cadmium, cobalt, copper, chromium, nickel, manganese, zinc, etc., in the acid region below pH 7 permits reduction of metal solubility to values which are orders of magnitude lower than are attainable by hydroxide precipitation. There are advantages to using combinations of sulfide and carbonate for precipitation with nickel, for example.[P-8, P-9] A merit of recovering metals as sulfides is that many metal refining operations, notably for copper, nickel, zinc, etc., are designed for processing sulfide ores.

There are some very specific precipitation procedures, such as the use of soluble barium salts to precipitate chromium as barium chromate for chromium recovery from metal finishing industry effluents.[P-10, P-11] There are also nonspecific, but efficient, precipitation processes which involve formation of insoluble metals by chemical reduction with sodium borohydride.[P-12, P-31, P-54, P-104] The cost of this latter process probably limits its application to precious metals.

Organic reactants have also been used to precipitate metals in solution by formation of insoluble products. Examples that can be cited are thiourea and thioacetamide for cadmium, copper, lead, chromium, titanium, molybdenum, and zinc,[P-99, P-120] dibromo-oxime for copper, cobalt, chromium, manganese, lead, and zinc;[P-121, P-122] and dialkyl dithiocarbamates for copper, iron, and zinc.[P-123] Complex formation with starch xanthates has been used to remove soluble metals from waste effluents.[P-97, P-101] Efficient recovery of the recovered metals from the insoluble metal-organic complex is essential for any practical process to be used for metal recovery so that poor regenerability and high initial cost would discourage use of most organic precipitants.

Coprecipitation of iron with other metal hydroxides to form ferrites has been applied to metal removal from waste effluents.[P-124, P-125] Formation of magnetic ferrites from such coprecipitates provides a further opportunity for a separation process (see Chapter 6.1, "Magnetic Separations").

The use of phosphate precipitants for trivalent metals such as aluminum, chromium, and iron provides an opportunity for separation of these metals as contaminants from divalent metals such as cadmium, copper, cobalt, nickel, zinc, etc., in mixed-metal waste solutions.[P-110, P-114]

Effective selective precipitation of metals such as cobalt, copper, nickel, and zinc as oxalates in a moderately acid to neutral pH range also provides an opportunity for separation of these metals as a chemical salt with market potential.[P-103, P-119, P-116]

Jarosite precipitation as mixed hydroxides from sulfate solutions for metal separations with arsenic and copper;[P-105] gold, silver, copper, cadmium, and lead;[P-106] arsenic, iron and lead;[P-111] and silver, lead, and zinc[P-112] have been reported to provide useful concentrations of these metals from various hydrometallurgical waste systems.

Finally, success in inhibiting precipitation in a moderately acid pH range of copper and nickel hydroxides in the presence of low concentrations of anionic surfactants such as ethylene diamine tetraacetic acid permits concentration of these metals in solution accompanied by precipitation separation of contaminant metals such as iron or aluminum.[P-102]

References

P-1.   Yost, K. J. and D. R. Masarik. "A Study of Chemical Destruct Waste Treatment Systems in the Electroplating Industry," *Plating Surf. Finish.* 64:(1977), p. 35.

P-2.   Brantner, K. A. and E. J. Cichon. "Heavy Metals Removal: Comparison of Alternate Precipitation Process," Proceedings of the 13th Mid-Atlantic Waste Conference (1981), p. 43.

P-3.   Bhattacharya, D. et al. "Separation of Toxic Heavy Metal Sulfides by Precipitation." *Sep. Sci. Technol.* 14:441 (1979).

P-4.   Robinson, A. K. and J. C. Sum. "Sulfide Precipitation of Heavy Metals." EPA 600/2–80–139; PB 80–225725 (1980).

P-5.   Yeligar, M. B. et al. "Treatment of Metal Finishing Wastes by Use of Ferrous Sulfide." EPA 600/2081–142 (1981).

P-6.   Higgins, T. E. and S. G. Termaath. "Treatment of Plating Waste Waters by Ferrous Reductions, Sulfide Precipitation, Coagulation and Upflow Filtration," in *Proceedings of the 36th Purdue Industrial Waste Conference,* J. B. Bell, Ed. (Ann Arbor, MI: Ann Arbor Science Books, 1983), p. 462.

P-7.   Gadd, R. K. and A. C. Sund-Hagelberg. U.S. Patent 4503017 (1985).

P-8.   McNally, S. and L. Benefield. "Nickel Removal from a Synthetic Nickel-Plating Waste Water Using Sulfide and Carbonate for Precipitation and Coprecipitation," *Sep. Sci. Technol.* 19:191 (1984).

P-9.   McNally, S. et al. "Nickel Removal from a Synthetic and Actual Nickel Plating Waste Water Using Sulfide and Carbonate for Precipitation and Coprecipitation," in *Proceedings of the 39th Purdue Industrial Waste Conference,* (Boston: Butterworth Publishers [Ann Arbor Science Books]; 1985), p. 81.

P-10.   Fadgen, T. J. "Precipitation of Sodium Dichromate with Barium," *Sewage Ind. Wastes* 24:1101 (1952).

P-11.   Fadgen, T. J. "Precipitation of Sodium Dichromate with Barium," *Sewage Ind. Wastes* 27:206 (1955).

P-12.   Jula, R. F. "Simplifying Heavy Metal Recovery with Sodium Borohydride," *Chem. Eng. Prog.*, Feb., 123 (1975).

P-13.   Tamai, Y., T. Okabe, and H. Ishii. "Recovery of Nickel from Residues." German Patent DE 2808263 (1978).

P-14.   Faul, W. and B. Kastening. "Precipitation of Metals from an Aqueous Solution Containing Metal Ions." German Patent DE 2717368 (1978).

P-15.   Badzynski, M., A. Brychczynski, L. Kopinski, R. Piechocki, E. Sobczak, and J. Wielunski. "Recovery of Nickel from Waste Baths and Slimes." Polish Patent PL 86221 (1976).

P-16.   O'Dell, C. G. "Reclamation of Metals from Plating Wastes," *Inst. Metall.* [*Course Vol.*], Ser. 3 (London), 8, 63–66, 112–113 (1977).

P-17.   Krochmal, J. "Pure Nickel Sulfate from Electrolyte for the Electrolytic Refining of Copper." Polish Patent PL 84609 (1977).

P-18.   Fukuoka, Y. and W. Abe. "Treatment of Plating Waste Solutions." *Jpn. Kokai Tokkyo Koho*, Japanese Patent JP 53132470 (1978).

P-19.   Bhattacharyya, D., A. B. Jumawan, Jr., and R. B. Grieves. "Separation of Toxic Heavy Metals by Sulfide Precipitation," *Sep. Sci. Technol.* 14(5):441–452 (1979).

P-20.   Katsura, T. and I. Nagata. "Treatment of Wastewater Containing High Zinc Content." *Jpn. Kokai Tokkyo Koho*, Japanese Patent JP 54041551 (1979).

P-21.   Novak, J. T., J. Holroyd, L. Pattengill, and M. M. Ghosh. "Optimum Dewatering and Metal Recovery of Metal Plating Waste Sludges," Report, CEEDO-TR-78–15; Order No. AD-A059957 avail. NTIS from: *Gov. Rep. Announce. Index (U.S.)* 1979, 79(3):108 (1978).

P-22.   Rolff, R. and H. J. Ehrich. "Apparatus and Methods for Recovery of Metals and Other Valuable Substances Present from Wastewater of Apparatus for Chemical Surface Treatment." German Patent DE 2747562 (1979).

P-23. Mikhnev, A. D., A. P. Serikov, A. I. Elesin, and L. A. Pshenichnikova. "Composition of Lime-Sulfur Decoctions," *Izv. Vyssh. Uchebn. Zaved., Tsvetn. Metall.* (1):127–128 (1980).

P-24. Puscasiu, M. G., D. V. Stanescu, A. M. M. Binder, P. S. Boian, V. Florea, E. I. P. Jimon, and M. Gaber. "Recovery of Nickel from Stainless Steel Scrap Metal" Romanian Patent RO 66600 (1979).

P-25. Wittman, E. "Circulating Rinsing System for Direct Recovery of Electrolytes from Surface Treating Baths." German Patent DE 3014847 (1980).

P-26. Popea, F. M., A. Boldijar, V. M. Danciu, and L. I. Patron. "Chemical Purification of Wastewaters Containing Nickel." Romanian Patent RO 75–84250 (1979).

P-27. Pazderka, J. "Equipment for Disposal and Regeneration of Wastewater," *Povrchove Upravy* 20(1):12–14 (1980).

P-28. Volcovinschi, G. and M. Keseru. "Recovery of Metal Compounds from Acid Wastewaters." Romanian Patent RO 67816 (1980).

P-29. White, W. W. "Process for Recovering Metals from Waste Solutions," *Res. Discl.* 209:362 (1981).

P-30. Walter, H., H. D. Reiterer, E. Eichberger, and G. Lazar. "Recovering Metals from Metallic Salt Solutions." European Patent EP 38322 (1981).

P-31. Cook, M. M. and J. A. Lander. "Sodium Borohydride Controls Heavy Metal Discharge," *Pollut. Eng.* 13(12):36–38 (1981).

P-32. Barna, V., I. Constantinescu, D. Dalalau, T. Dragulescu, and L. M. Bogdan. "Recovery of Metal Components of Hard Alloys." Romanian Patent RO 66138 (1980).

P-33. Sumitomo Metal Mining Co., Ltd. "Recovery of Gold from the Anodic Sludge of Electrolytic Copper Refining." Belgian Patent BE 891130 (1982).

P-34. Christenson, C. P., G. M. Mcnamee, and R. H. Delaune. "Removing Copper from Waste Complex Solutions of Copper Aluminum Halides." Brazilian Patent BR 8100704 (1982).

P-35. Vainshtein, I. A., L. D. Klenysheva, L. L. Polishchuk, T. V. Matveeva, G. I. Khaustov, I. Y. Korobchkin, V. A. Tarasenko, and I. F. Lavretskaya. "Processing Fluorine-Containing Solutions." U.S.S.R. Patent SU 971817 (1982).

P-36. Drule, L. M., I. Costea, and L. Deac. "Recovery of Copper, Zinc, and Nickel from German Silver Wastes." Romanian Patent RO 79679 (1982).

P-37. Jenke, D. R. and F. E. Diebold. "Recovery of Valuable Metals from Acid Mine Drainage by Selective Titration," *Water Res.* 17(11):1585–1590 (1983).

P-38. Hans, W., H. D. Reiterer, E. Eichberger, and G. Lazar. "Recovering of Metals from Solutions of Metal Salts." U.S. Patent US 4406696 (1983).

P-39. Perte, E., G. Marcu, O. Ceuca, O. Crucin, L. Pacuraru, C. Ghiara, and T. Budiu. "Recovery of Palladium from Silver and Copper Alloy Wastes on a Pinchbeck Support." Romanian Patent RO 79963 (1982).

P-40. Krotz, K. J. "Treating Spent Fluids to Recover Copper and Copper Oxide." U.S. Patent US 4428773 (1984).

P-41. McAnally, S., L. Benefield, and R. B. Reed. "Nickel Removal from a Synthetic Nickel-Plating Wastewater Using Sulfide and Carbonate for Precipitation and Coprecipitation," *Sep. Sci. Technol.* 19(2–3):191–217 (1984).

P-42. Morimoto, A. "Treatment of Waste Liquids Containing Copper." *Jpn. Kokai Tokkyo Koho*, Japanese Patent JP 59010390 (1984).

P-43. Leroy, J. B. "Copper Recovery in the Treatment of Wastewater from Printing Works: Example of the

Limay Plant," in *Recycl. Int.: Recovery Energy Mater. Residues Waste* [Contrib.−Int. Recycl. Congr. (IRC)], Thome-Kozmiensky, K.J., Ed. (Berlin: E. Freitag-Verlag Umwelttech, 1982), 592–595.

P-44.   Schmack, P., H. Morgenstern, E. Raphael, L. Hiller, and H. Ungaenz. "Recovery of Copper from Waste-waters of Chemical Purification Processes in Power Stations." East German Patent DD 209170 (1984).

P-45.   Cocheci, V., L. Haias, A. Martin, and V. Scarlat. "Purification of Wastewater from Nonferrous Metal-lurgy by Precipitation with Calcium Polysulfide and Simultaneous Recovery of Useful Metals," *Bul. Stiint. Teh. Inst. Politeh., Timisoara, Ser. Chim.* 28(1–2):1–8 (1983).

P-46.   Hepp, W., G. Fessel, F. Kleine, J. Von Lehe, P. Friedemann, and F. Gensch. "Treatment of Waste-water from Copper Electroplating of Welding Wire." East German Patent DD 211106 (1984).

P-47.   Peters, R. W., Y. Ku, and D. Bhattacharyya. "Crystal Size Distribution of Sulfide Precipitation of Heavy Metals," *Process Technol. Proc. 2 (Ind. Cryst.)* 111–123 (1984).

P-48.   Sumitomo Metal Mining Co., Ltd. "Ferric Chloride Etchant Regeneration." *Jpn. Kokai Tokkyo Koho*, Japanese Patent JP 59200764 (1984).

P-49.   Isabaev, S. M., E. G. Mil'ke, A. N. Polukarov, K. Kuzgibekova, and K. Zhumashev. "Processing of Copper Dross by Melting." Russian Patent SU 1134616 (1985).

P-50.   Hirano, K., T. Yokoi, K. Iguchi, and T. Tanabe. "Study on Electrolytic Recovery of Nickel from Nickel Electroplating Industrial Wastewater by the Dia-phragm Electrolytic Process," *Kenkyu Hokoku − Kanagawa-ken Kogai Senta* 6:22–27 (1984).

P-51.   Nemes, I., G. Faur, I. Costea, and I. Vida. "Recovery of Useful Elements from Wastes from Semiconductor Device Manufacture." Romanian Patent RO 83221 (1984).

P-52.   Wolski, T. and J. Glinski. "The Utilization of Pickling Effluents in Agriculture," *Pol. Tech. Rev.* (5–6):16–17 (1984).

P-53.   Wolski, T. and J. Glinski. "Solid Industrial Wastes as a Source of Microelements for Agriculture," *Pol. Tech. Rev.* (5–6):18–19 (1984).

P-54.   Tuznik, F. S. and A. A. Lis. "Environmental Protection and Raw Materials Recovery with FKJA/LAFT System in the Metal Finishing Industry," Proceedings of the 11th World Congress Met. Finish., Tel Aviv (1984), pp. 297–305.

P-55.   Rodriguez, D., R. Schemel, and R. Salazar. "Precipitation or Recovery of Vanadium from Liquids." German Patent DE 3509372 (1985).

P-56.   Slawski, K., L. Wedzicha, Z. Mromlinska, and A. Waligora. "Recovery of Copper and Zinc from Effluents from the Brass Electroplating," *Rudy Met. Niezelaz.* 30(5):191–193 (1985).

P-57.   Sato, H., M. Kataoka, R. Koshio, and I. Nakajima. "Copper Recovery from Spent Electroless Copper Bath." *Jpn. Kokai Tokkyo Koho*, Japanese Patent 60187634 (1985).

P-58.   Friedman, R. H. "Recovering Metals from Waste." U.S. Patent US 4579721 (1986).

P-59.   Constantinescu, I. and L. Oproiu. "Recovery of Tungsten, Copper, and Nickel from Waste Sintered Electric Contacts." Romanian Patent RO 86886 (1985).

P-60.   Slawski, K. and Z. Mromlinska. "Recovery of Zinc from Wastes of Lead Soldering of Copper Elements," *Rudy Met. Niezelaz.* 30(8):296–298 (1985).

P-61.   Slawski, K., L. Wedzicha, and L. Zurowska. "Nickel Recovery from Galvanic Wastes," *Rudy Met. Niezelaz.*, 31(2):42–44 (1986).

P-62.   Mitulla, K., J. Hambrecht, S. Marquardt, H. Brandt, B. Schmitt, H. Gausepohl, P. Siebel, and H. Dreher. "Recovery and Reuse of Amines and Metal Components in the Preparation of Polyoxyphenylenes." German Patent DE 3522470 (1987).

P-63. Lehmann, R., D. Polanco, R. Brockl, and R. Schemel. "Concepts for Heavy Metal Handling in Heavy Oil Upgrading Processes," in 3rd International Conference Heavy Crude Tar Sands, Vol. 4 (New York: UNITAR/UNDP Inf. Cent. Heavy Crude Tar Sands, 1985), pp. 2053–2072.

P-64. Wiegand, V. "The Hematite Process in Zinc Electrolysis as an Example of Waste Utilization," *Schriftenr. GDMB*, 47(*Abfallst. Nichteisen-Metall.*), 191–203 (1986).

P-65. Abe, T., T. Yoshihara, T. Sano, and T. Mimuro. "Combined Treatment of Spent Acid and Metal Oxide-Metal Mixture." *Jpn. Kokai Tokkyo Koho*, Japanese Patent JP 61284538 (1986).

P-66. Van Dijk, J. C., P. J. De Moel, and M. Scholler. "Recovery of Heavy Metals in the Electroplating Industry," *PT-Procestech.*, 49(1):33–37 (1986).

P-67. Zhang, Y., J. Chen, R. Zhang, and W. Gao. "Recovery of Valuable Metals from Industrial Waste." *Faming Zhuanli Shenqing Gongkai Shuomingshu*, Chinese Patent CN 85100731 (1986).

P-68. Frankard, J. M. and O. V. Broch. "Separation and Recovery of Reusable Heavy Metal Hydroxides from Metal Finishing Wastewaters." U.S. Patent US 4680126 (1987).

P-69. Charewicz, W. A., T. Chmielewski, B. Kolodziej, and J. Wodka. "Hydrometallurgical Recovery of Nickel from Spent Catalysts," *Rudy Met. Niezelaz.* 32(2):61–65 (1987).

P-70. Sato, H., M. Kataoka, T. Yoshida, and H. Yoshino. "Copper Recovery from Spent Chemical Coating Solution." *Jpn. Kokai Tokkyo Koho*, Japanese Patent JP 62202034 (1987).

P-71. Antalik, B. and A. Czeke. "Salt Recovery from Natural Waters and Wastewaters." *Hung. Teljes*, Hungarian Patent HU 42409 (1987).

P-72. Schoeller, M., J. C. Van Dijk, and D. Wilms. "Recovery of Heavy Metals by Crystallization," *Met. Finish.* 85(1):31–34 (1987).

P-73. Schoeller, M., J. C. Van Dijk, and D. Wilms. "Recovery of Heavy Metals by Crystallization in the Pellet Reactor," in *Proceedings of the 2nd European Conference on Environmental Technology*, K. J. A. De Walls and W. J. van den Brink, Eds. (Dordrecht, The Netherlands: Martinus Nijhoff, 1987), pp. 204–303.

P-74. Heimhard, H. J., H. Werner, and J. Allef. "Treatment of Metal Industry Wastewaters to Recover Metals." German Patent DE 3611448 (1987).

P-75. Masuda, Y. "Recovery of Tin from Scraps." *Jpn. Kokai Tokkyo Koho*, Japanese Patent JP 62250133 (1987).

P-76. Lindquist, B. "Heavy Metals Precipitation from Smelter Waste Water and Obtainable Sludge Qualities," in *Management of Hazardous Toxic Wastes Process Ind., [Int. Congr.]*, S. T. Kolaczkowski and B. D. Crittenden, Eds. (London: Elsevier Applied Science Publishers, 1987), pp. 310–327.

P-77. Brooks, C. S. "Nickel Metal Recovery from Metal Finishing Industry Wastes," *Proceedings of the 42nd Purdue Industrial Waste Conference*, Volume Date 1987 (Chelsea, MI: Lewis Publishers, Inc., 1988), pp. 847–852.

P-78. Jellinek, H. H. G. and Ming Dean Luh. "A Novel Method of Metal Ion Removal and Recovery from Water by Complex Formation with Polyelectrolytes," *J. Polymer Sci.*, Part A-1, 7:2445–2449 (1969).

P-79. Haikawa, K. "Treatment of Waste Solution for Non-electrolyte Copper Precipitation." *Jpn. Kokai Tokkyo Koho*, Japanese Patent 63134639 (1988).

P-80. Von Roepenack, A., W. Boehmer, G. Smykalla, and V. Wiegand. "Processing of Residues from Hydrometallurgical Recovery of Zinc." German Patent DE 3634359 (1988).

P-81. Wolski, T. "Recovery of Copper from Spent Etching Solutions." Polish Patent PL 133852 (1987).

P-82.   Shukla, A., P. N. Maheshwari, and A. K. Vasishtha. "Reclamation of Nickel from Spent Nickel Catalyst," *J. Am. Oil Chem. Soc.* 65(11):1816–1823 (1988).

P-83.   Ward, V. C. "Meeting Environmental Standards When Recovering Metals from Spent Catalyst," *J. Occup. Med.* 41(1):54–55 (1989).

P-84.   Martinek, P. and I. Rousar. "Processing of Nickel Alloy Scrap and Waste." Czechoslovakian Patent CS 255054 (1988).

P-85.   Mikhailovskii, V. L., V. E. Ternovtsev, Y. S. Sergeev, L. A. Gergalov, V. P. Dubrovskii, S. V. Sokolov, and O. I. Kochetov. "Recovery of Copper from Solutions." U.S.S.R. Patent SU 1475950 (1989).

P-86.   Reuter, R. "Metal Recovery from Hydroxide Sludges, for Example, from Galvanization." German Patent DE 3729913 (1989).

P-87.   Ilangovan, R., G. Govindarajan, S. Ramamurthi, and N. V. Parthasaradhy. "Treatment and Recovery of Fluoroborate Spent Plating Baths," *Trans. Soc. Adv. Electrochem. Sci. Technol.* 24(2):181–185 (1989).

P-88.   Sheng, S. "Recovery of Copper from Etching Solution Containing Ferric Chloride for Printed Circuit Boards," *Diandu Yu Huanbao* 9(1):44–46 (1989).

P-89.   Leggett, D. J. and J. G. Courtwright. "Recovery of Metals from Solution Containing Chelants." U.S. Patent US 4846978 (1989).

P-90.   Sefton, V. B., R. Fox, and W. P. Lorenz. "Recovery of Metal Values from Spent Petroleum Refining Catalyst." U.S. Patent US 4861565 (1989).

P-91.   Schmidt, J. and W. Bamberg. "Detoxification of Solutions of Copper Cyanide Complexes with Recovery of Copper." East German Patent DD 270322 (1989).

P-92.   He, Y. "Silver Recovery from Wastewater Containing Silver Ions," *Huanjing Baohu (Beijing)* (5):20 (1989).

P-93.   Ni, D. "Treatment and Utilization of Industrial Wastewater in the Dexin Copper Mine," *Huanjing Kexue* 10(6):34–38 (1989).

P-94. Nakazawa, H., T. Miyamoto, and H. Sato. "Selective Precipitation of Iron Ion from a Hydrochloric Waste Etch," *Shigen to Sozai* 106(3):145–149 (1990).

P-95. Csiriny, G. and C. Nemeth. "Neutralization and Recovery of Copper from Spent Solutions Containing Amines." *Hung. Teljes,* Hungarian Patent HU 49914 (1989).

P-96. McFadden, F., L. Benefield, and R. B. Reed. "Nickel Removal from Nickel Plating Wastewater, Using Iron, Carbonate and Polymers for Precipitation and Coprecipitation," in *Proceedings of the 42nd Industrial Waste Conference, Purdue University* (Boston: Butterworth Publishers, Inc., 1985), p. 417.

P-97. Bricka, R. M. and M. J. Cullinane, Jr. "Comparative Evaluation of Heavy Metal Immobilization Using Hydroxide and Xanthate Precipitation," in *Proceedings of the 42nd Industrial Waste Conference, Purdue University* (Chelsea, MI: Lewis Publishing Co., 1988), p. 809.

P-98. Ying, W. C., R. R. Bonk, and M. E. Tucker. "Precipitation Treatment of Spent Electroless Nickel Plating Baths," in *Proceedings of the 42nd Industrial Waste Conference, Purdue University* (Chelsea, MI: Lewis Publishing Co., 1988), p. 831.

P-99. Nelson, J. H., J. L. Hendrix, and E. Miloszvljevic. "Use of Thiourea and Thioacetamide for Separation and Recovery of Heavy Metals from Mineral Treatment Waste Water," in *Metals Speciation, Separation and Recovery,* J. W. Patterson and R. Passino, Eds. (Chelsea, MI: Lewis Publishing Co., 1987), p. 119.

P-100. Peters, R. W. and Y. Ku. "The Effect of Citrate, a Weak Chelating Agent, on the Removal of Heavy Metals by Sulfide Precipitation," in *Metals Speciation, Separation and Recovery,* J. W. Patterson and R. Passino, Eds. (Chelsea, MI: Lewis Publishing Co., 1987), p. 147.

P-101. Tiravanti, G., A. C. Di Pinto, G. Macchi, D. Marani, M. Santow, and Y. Wang. "Heavy Metals Removal:

Pilot Scale Research on the Advanced Mexico Precipitation Process," in *Metals Speciation, Separation and Recovery*, J. W. Patterson and R. Passino, Eds. (Chelsea, MI: Lewis Publishing Co., 1987), p. 665.

P-102. Brooks, C. S. "Metal Recovery by Selective Precipitation. I. Hydroxide Precipitation," *Met. Finish.* 88(11):21 (1990).

P-103. Brooks, C. S. "Metal Recovery by Selective Precipitation. II. Oxalate Precipitation," *Met. Finish.* 88(12):15 (1990).

P-104. Duncan, R. N. and J. R. Zickgraf. "One Way to Treat Spent EN Baths," *Prod. Finish.* 46(4):54 (1982).

P-105. Takahashi, N. and H. Toda. "Copper-Arsenic Compound." German Patent No. 3048404 (1982).

P-106. Pammenter, R. V. and C. J. Haigh. "Improved Metal Recovery with the Low-Contaminant Jarosite Process," in *Extr. Metall. '81, Pap. Symp.* (London: Inst. Min. Metall., 1981), pp. 379–392.

P-107. Zembura, Z., L. Burzynska, W. Glodzinska, and B. Cholocinska. "Recovery of Nickel and Cobalt from Waste Materials," *Rudy Met. Niezelaz.* 27(8):372–375 (1982).

P-108. Osaka Prefecture. "Nickel Recovery from Waste Chemical Coating Solutions." Japanese Patent No. 59185770 (1984).

P-109. Yamasaki, K., T. Morikawa, and S. Eguchi. "Treatment of Spent Copper Electroless Coating Baths by Tartrate Salt Precipitation," *Osaka-furitsu Kogyo Gijutsu Kenkyusho Hokoku* (86):1–6 (1985).

P-110. Dahnke, D. R., L. G. Twidwell, B. W. Arthur, and S. M. Nordwick. "Selective Recovery of Metal Values from Electroplating Sludges by the Phosphate Process." *Annual Technical Conference Proceedings— American Electroplating Society*, 73rd (Sess. C) (1986), Paper C-3.

P-111. Plasket, R. P. and G. M. Dunn. "Iron Rejection and Impurity Removal from Nickel Leach Liquor at Impala Platinum Limited," in *Iron Control Hydrometall.*

[*Int. Symp.*], J. E. Dutrizac and A. J. Monhemius, Eds. (Chichester: Ellis Horwood, 1986), pp. 695–718.

P-112. Peters, M. A. and W. W. Hazen. "Process for Recovering Metal Values from Ferrite Wastes." U.S. Patent No. 8803912 (1988).

P-113. Meloyan, R. G., R. S. Edilyan, and A. A. Mushegyan. "Recovery of Copper from Industrial and Recycling Solutions," *Tsvetn. Met. (Moscow)* (7):57 (1988).

P-114. Ying, W. C. and R. R. Bonk. "Treatment of Electroless Nickel Baths." U.S. Patent US 4789484 (1988).

P-115. Brooks, C. S. "Treatment and Metal Recovery for Electroless Metal Plating Wastes," *Proceedings of the 43rd Purdue Industrial Waste Conference*, Volume Date 1988 (Chelsea: MI: Lewis Publishers, Inc., 1989), pp. 721–726.

P-116. Brooks, C. S. "Metal Recovery from Electroless Plating Wastes," *Met. Finish.* 87(5):33–36 (1989).

P-117. Slawski, K., Z. Mromlinska, L. Wedzicha, and A. Waligora. "Recycling of Materials When Electroplating in Cyanidic Electrolytes," *Metalloberflaeche* 44(1):13–14 (1990).

P-118. Ctvrtnicek, A., I. Capova, M. Zapletalek, K. Vurm, and J. Pucherna. "Hydrometallurgical Processing of Waste Matte Containing Lead." Czechoslovakian Patent 259706. (1989).

P-119. Mallory, E. C. Jr. *Trace Inorganics in Water, Adv. Chem. Ser.* 73:281–295 (1968).

P-120. Fishman, M. J. and E. C. Mallory, Jr. *J. Water Pollut. Control Fed.* 40:R67–R71 (1968).

P-121. Riley, J. P. and G. Topping. *Anal. Chim. Acta* 44:234–236 (1969).

P-122. Pan, Y. and G. Ouyang. "Treatment of Copper-Containing Acidic Wastewater and Recovery of Copper," *Huanjing Baohu (Beijing)* (11):18–19 (1986).

P-123. Zievers, J. F. "Emerging Technologies for Waste Water Treatment," *Ind. Finish.* 15:22 (1975).

P-124. Okamoto, S. "Iron Hydroxides as Magnetic Scavengers," *I.E.E.E. Trans. Magn.* 10:923 (1974).

P-125. Okuda, T. et al. "Removal of Heavy Metals from Wastewater by Ferrite Co-precipitation," *Filt. Sep.* 12:472 (1975).

6

Separation of Solid Wastes

Metal separation processes directly applicable to solid wastes consist of biological processes, flotation, magnetic separations, pyrometallurgical processes, and solvent partition. These processes are most successful for metal wastes with minimal amounts of secondary contaminant metals and here provide potential opportunities for dewatering and/or concentration if not separation.

6.1 BIOLOGICAL SEPARATIONS

Biological processes can also play a role in the separation of metals. One type of process arises from the solubilization effect of organisms, such as certain bacteria, on minerals and various solid wastes. The leaching of copper sulfide ores with *Thiobacillus ferrooxidans* has been used for some time as a commercial process for copper recovery. There are a number of other biological systems, such as *Sulfolobus, Pseudomonas, Spirogyra, Oscillatoria, Rhizoclonium, Chara,* and *Synechococcus,* that also have received attention as offering promise for the extraction of metals such as iron, cadmium, copper, chromium, mercury, nickel, manganese, lead, molybdenum, selenium, uranium, tin, cesium, radium, and aluminum from ores or solid wastes.

Another type of separation process that has been considered is the accumulation by adsorption-ion exchange of metal cations in aqueous solution onto organic substrates,

Table 6.1 Metal Separations Using Biological Agents

Waste system	Metals	Biological agent	Mode of metal action	Ref.
Ores	Fe	*Thiobacillus ferrooxidans*	Leach	B-2
Polluted water	Mo, Ra, Se, V	*Spirogyra, Oscillatoria Rhizoclonium + Chara*	Accumulate	B-3
Metal finishing waste	Cd, Cu, Pb	*Chlorella pyrenoidosa*	Accumulate	B-5
Sewage sludge	Cd, Cu, Ni, Zn	*T. ferroxidans*	Leach	B-7
Wastewater or leachate	Cu	*Desulfavibrio vulgaris*	Accumulate	B-11
Nuclear wastewater	^{241}Am, Ce, Co, Cs, Cu, ^{242}Pu, ^{134}Cs, ^{85}Sr, Sr, Zr	*Bacillus subtilisis*	Accumulate	B-12
Metal wastewater or leachate	Cu, U	*T. ferrooxidans, T. thiooxidans, Leptospirillam ferrooxidans*	Leach	B-13
Metal wastewater	Cd, Cu, Ni, Pb, Zn	*Penicillum, Cladosporium*	Accumulate	B-14
Metal wastewater	Ag, Al, Cd, Cr, Co, Cu, Hg, Ni, Pb, Zn	*Chlorella pyrenoidosa, Spirulina*	Accumulate	B-15
Ores	Ag, Cd, Cu, Pb, Zn	*T. ferrooxidans, Chlamydomonas reinhardtii*	Leach	B-16

notably algae such as *Chlorella pyrenoidosa, Spirulina, Penicillium,* and *Cladosporium.* Accumulation processes on organic substrates offer promise for removal of soluble metals from liquid wastes, and it has been known for some time that passage of polluted waters through fresh and marine wetlands provides quite a significant amount of cleanup.

One interesting example[B-12] for purification of nuclear industry wastes is the use of an aerobic recovery of ^{242}Pu, ^{241}Am, ^{134}Cs, ^{85}Sr, ^{60}Cu, U, Ru, Sr, Co, Cs, Ce, and Zr with *Bacillus subtilis* with subsequent recovery of the metals by magnetic separation. In addition, the anaerobic action of *Desulfovibrio* in a sulfate-lactate culture collects copper sulfate for subsequent magnetic separation. Another interesting biological separation process involves precipitation of copper or other metals from leach solutions or wastewaters as sulfides with *Desulfovibrio vulgaris,* followed by oxidation of the resultant H_2S with *Chromatium vinosum* bacteria.[B-11]

Practical success with the biological separation processes requires coping with large volume reaction systems and the patience needed for low-temperature reaction kinetics. A summary is provided in Table 6.1 of the various waste and mineral systems that have been examined, the types of metals involved, the identified organisms, and an indication of whether or not the process is appropriate for leaching or accumulation separations.

References

B-1.   Kelley, D. P. et al. "Microbial Methods for the Extraction and Recovery of Metals," in *Microbial Technology: Current Status, Future Prospects,* A. T. Bull et al., Eds., Society for General Microbiology Symposium No. 29 (Cambridge: Cambridge University Press, 1979).

B-2.   Monroe, D. "Microbial Metal Mining," *Am. Biotechnol. Lab.* 3:10(1985).

B-3.   Brierley, C. L. "Microbiological Mining," *Sci. Am.* (August 1982), p. 44.

B-4. Short, H. and G. Parkinson. "Rx for Leaching Copper: Go the Biological Way," *Chem. Eng.* (July 11, 1983), p. 26.

B-5. Sloan, F. J., A. R. Abernathy, and J. C. Jennett. "Algae, Ion Exchange and Metal Finishing Wastes," *39th Purdue Industrial Waste Conference* (Boston: Butterworth Publishers, 1985), p. 537.

B-6. Clyde, R. A. "Method for Treating Waste Fluid with Bacteria." U.S. Patent 4530763 (1985).

B-7. Wong, L. and J. G. Heur. "Biological Removal and Chemical Recovery of Metals from Sludges," in *39th Purdue Industrial Waste Conference* (Boston: Butterworth Publishers, 1985), p. 515.

B-8. Kunicki-Goldfinger, W., M. Ostrowski, and A. Lejeczak. "Microbiological Leaching of Copper from Covellite and Post-Flotation Wastes in an Alkaline Environment," *Fizykochem. Probl. Mineralurgii* 13:165–172 (1981).

B-9. Almasan, B., E. Constantinescu, S. Benea, V. Strat, and F. Burlacu. "Application of Copper-Leaching Techniques to Low-Grade Ores and Waste Dumps of the Altin Tepe Mine [Romania]," *Mine. Pet. Gaze.* 35(2):76–82 (1984).

B-10. Kataoka, M. and H. Sato. "Treatment of Chemical Plating Waste-Waters." Japanese Patent No. 60193584 (1985).

B-11. Imai, K. "Utilization of Sulfate-Reducing and Photolithotrophic Bacteria in Biohydrometallurgy," *Process Metall.* 4(Fundam. Appl. Biohydrometall.): 383–394 (1986).

B-12. Watson, J. H. P. and D. C. Ellwood. "Bacteria and a Magnet Separate Heavy Metals from Solutions," *I²-Procestechnologie* 5(11):21, 23–26 (1989).

B-13. Brierley, J. "Biotechnology for the Extractive Metals Industries," *J. Metals* 42(1):28 (1990).

B-14. Siegel, S. M., M. Galun, P. Keller, B. Z. Siegel, and E. Galun. "Fungal Biosorption: A Comparative Study of Metal Uptake by Penicillium and Cladosporium," in

Metals Segregation, Separation and Recovery, J. W. Patterson and R. Rassino, Eds. (Chelsea, MI: Lewis Publishers, Inc., 1987), p. 339.

B-15. Greene, B., R. McPherson, and D. Darnall. "Algal Sorbents for Selective Metal Ion Recovery," in *Metals Segregation, Separation and Recovery*, J. W. Patterson and R. Rassino, Eds. (Chelsea, MI: Lewis Publishers, Inc., 1987), p. 315.

B-16. Frenay, J., J. Remacle, M. Cline, R. Matagne, J. M. Collard, and J. Wiertz. "Microbial Recovery of Metals from Low Grade Materials," in *Recycle and Secondary Recovery of Metals*, P. R. Taylor, H. Y. Sohn, and N. Jarrett, Eds. (Warrendale, PA: Metallurgical Society, 1985), p. 275.

6.2 FLOTATION

Separation and concentration of soluble metals and insoluble metal hydroxides by flotation involves attachment to air bubbles. The nomenclature of flotation has become rather involved, but the classification proposed by Lemlich[F-1] will be used in the present discussion. No less than seven categories are included under the heading of "froth flotation," Two of these categories (ore flotation and precipitate flotation) are applicable to metals initially in an insoluble state and three (ion flotation, molecular flotation, and adsorbing colloid flotation) are applicable to metals initially in a soluble state. Some relevant published results are discussed below.

An extensive literature on ore flotation[F-2 to F-11, F-54, F-61, F-64] exists. Flotation of oxides is not as common a commercial practice as that for sulfides, but considerable research has been published. Anionic agents, such as potassium octylhydroxamate, have been used for flotation of ferric oxide.[F-1, F-12, F-13] Cationic agents, such as dodecylamine, dodecyl ammonium salts, and alkyl pyridinium salts, have been evaluated for flotation of oxide minerals, such as corundum, hematite, and geothite.[F-6] Anionic agents, such as carboxylates as well as alkane sul-

Table 6.2 Metal Recovery by Flotation

Waste system	Metals	Flotation agent	Flotation mode	Metal separation efficiency %	Ref.
Cassiterite	Sn	Salicylaldehyde	Foam flotation		F-8
Scheelite	W		Foam flotation		F-10
Fe_2O_3	Fe	Octylhydroxamate	Foam flotation		F-12
Metal solution	Cu	Ethyl(hexadecyldimethyl) ammonium bromide	Ion flotation		F-17
Metal solution	Pb	Monoalkyl phosphate	Ion flotation		F-18
Metal solution	Hg	Amidoxime surfactant	Ion flotation		F-19
Metal solution	Cu, Cr, Ni, Pb, Zn	Ethyl (hexadecyl dimethyl) ammonium bromide + Na dodecyl benzene sulfonate	Micro gas flotation		F-24
Coal/phosphate slime		As above	Micro gas flotation		F-25
Metal solution	Cu	Dodecylamine N, N′, diacetic acid	Adsorption colloid + Fe(OH)₃ floc		F-26
Metal solution	Cd, Cu, Ni	4-Dodecyldiethylene triamine	Foam flotation		F-28
Metal solution	Cd, Hg	Hexadecyltrimethyl ammonium chloride	Adsorption colloid + Fe/Al hydroxide		F-30
Metal solution	Co, Cr, Cu, Ni, Zn	Na lauryl sulfate + Fe hydroxide floc	Adsorption colloid + Fe/Al hydroxide		F-31
Metal solution	Co, Cr, Ni	Hexadecyltrimethyl ammonium bromide + Na lauryl sulfate	"		F-32
Metal solution	Co, Cr, Cu, Ni, Pb, Zn	Na oleate + Na lauryl sulfate	"		F-33

Source	Metals	Reagent	Method	Results	Ref.
Metal solution	Ni	Na lauryl sulfate + dimethyl syloxime	Adsorption colloid with Ni hydroxide		F-60
Metal solution	Cu	Methyldodecyl benzyl trimethyl ammonium chloride	"		F-35
Metal solution	Cd, Cu, Hg, Zn	Na oleate + Na lauryl sulfate	"		F-36
Metal solution	Cu, Mn, Zn	Monolauryl phosphate	"		F-39
Seawater	Cu, Zn	Dodecylamine + $Fe(OH)_3$ floc	"	95	F-40
Ferromanganese nodules	Co, Cu, Mn, Ni	8 Hydroxyquinoline, dithiozone, ammonium 1-pyrrolidinedithiocarbamate	"		F-43
Wastewater	Ni, V, W	C_{16}–C_{18} carbolic acids	Ion flotation	92 W; 99 V; 95 Ni	F-46
Ethylene diamine solution	Cu, Ni	Na alkylarenesulfonate-resin acid soaps	Ion flotation		F-49
Metal solution	Co, Mn, Ni, V	N. decanol diethylenetriamine	Foam flotation	60 Ni	F-50
Wastewater	Cu, Mo, U, Zn	Dicresol dithiophosphate, Bu/Etxanthate	Foam flotation	98 Cu, 96 Mo, 92 U, 90 Zn	F-51
Wastewater	Cu	Laurylamine acetate	Foam flotation	>90 Cu	F-58
Wastewater	Cd, Cr, Cu, Fe, Ni, Zn	Dodecylbenzene sulfonate + metal hydroxide flocs	Adsorption colloid flotation	48–73 for other hydroxides 89 for $Cu(OH)_2$	F-59
Wastewater	Cu	Dodecyltrimethyl ammonium + tetradecyltrimethyl ammonium chloride	Foam flotation	>95 Cu	F-68

fates, sulfonates, and hydroxamates, usually in the C_8 to C_{18} range, have been examined for flotation of mineral oxides, such as alumina, quartz, hematite, pyrolusite, chromite, chrysocolla, and metal hydroxides of lead, iron, calcium, magnesium, manganese, and copper.[F-7, F-14] An anionic chelation agent such as salicylaldehyde has been used for flotation of cassiterite[F-8] and oleic acid has been applied to the flotation of ilmenite ore.[F-9] Scheelite has been selectively floated with mixtures of oleic and naphthienic acid.[F-10]

Precipitate flotation[F-2, F-5, F-11, F-26, F-29 F-30, F-32 to F-38, F-40, F-44, F-59, F-70 to F-72, F-74 to F-76] has been used to transform metal ions in aqueous solution to insoluble species amenable for attachment to air bubbles for separation by froth flotation. Attention has been directed to techniques for flotation of the fine particles less than 10μm, which are below the size desired for commercial ore flotation.[F-20 to F-25, F-76, F-77] These techniques include generation of microbubbles and the use of chemisorbing collectors, high-temperature flotation, carrier flotation, column flotation, oil flotation, liquid-liquid extraction, flocculation, electroflotation, vacuum or pressure release flotation, and precipitate flotation. Some of these techniques merit special consideration for their applicability to the fine particle sizes typical of waste metal oxide sludges.

Ion flotation has been used to separate cations of metals such as gold, silver, iron, chromium, copper, cobalt, and zinc,[F-17 to F-19, F-27 to F-29, F-37, F-38] but studies have been primarily experimental with limited reduction to commercial practice.

Adsorbing colloid flotation or floc foam flotation procedures have been developed for efficient removal of low concentrations of soluble metals in wastewaters,[F-31, F-33, F-36, F-38, F-60, F-70 to F-77] in seawater,[F-40 to F-44] and in acid-digested ferromanganese nodules.[F-43] Flotation can be used to concentrate insoluble metal hydroxides (copper, chromium, nickel, and zinc) in small particle sizes from aqueous slurries and suspensions. Bench-scale experimentation with actual and simulated waste metal hydroxides has produced promising results for nonselective flotation separations with sodium lauryl sulfate in combination with nonionic flocculants in the

6 to 8 pH range.[F-60] Highly selective flotation separations of metals such as nickel have been obtained using chelation reagents, such as dimethyl glyoxime, in combination with sodium lauryl sulfate.

A summary is provided in Table 6.2 (pages 118–119) of the various materials, metals, metal oxides, metal sulfides, etc.; the principal surfactants used to achieve attachment of the solid particles to the air bubbles; and the special features of the flotation systems for a wide variety of wastes.

References

F-1.   Lemlich, R., Ed. *Adsorptive Bubble Separation Techniques* (New York: Academic Press, 1972).

F-2.   Fuerstenau, M. C., Ed. *Flotation* – A.M. Gaudin Memorial Vols. 1 and 2 (New York: AIME, 1976).

F-3.   Gaudin, A. M. *Principles of Mineral Dressing* (New York: McGraw-Hill, 1939).

F-4.   Gaudin, A. M. *Flotation*. 2nd ed. (New York: McGraw-Hill, 1957).

F-5.   Fuerstenau, D. W. and Healy, T. M. *Adsorptive Bubble Separation Techniques*, R. Lemich, Ed. (New York: Academic Press, 1972), chap. 6.

F-6.   Smith, R. W. and S. Akhtar. "Cationic Flotation of Oxides and Silicates," in *Flotation* – A.M. Gaudin Memorial Vol. 1 (New York: AIME, 1976), pp. 87–116.

F-7.   Fuerstenau, D. W. and B. R. Palmer. "Anionic Flotation of Oxides and Silicates," in *Flotation* – A.M. Gaudin Memorial Vol. 1 (New York: AIME, 1976), pp. 148–196.

F-8.   Rinelli, G. et al. "Flotation of Cassiterite with Salicylaldehyde as a Collector," in *Flotation* – A.M. Gaudin Memorial Vol. 1 (New York: AIME, 1976), pp. 549–560.

F-9.   Perkins, E. J. and H. L. Shergold. "The Effect of Temperature on the Conditioning and Flotation of an Ilmemite Ore," in *Flotation* – A.M. Gaudin Memorial Vol. 1 (New York: AIME, 1976), pp. 561–579.

F-10. Vazquez, L. A., et al. "Selective Flotation of Scheelite," in *Flotation*—A.M. Gaudin Memorial Vol. 1 (New York: AIME, 1976), pp. 580–596.

F-11. Somasundaran, P. "Separation Using Foaming Techniques," *Sep. Sci.* 10:93 (1975).

F-12. Raghavan, S. and D. W. Fuerstenau. "Adsorption of Aqueous Octylhydroxamate on Ferric Oxide," *J. Colloid Interface Sci.* 50:319 (1975).

F-13. Eisenhauer, J. and E. Matijevic. "Interactions of Metal Hydrous Oxides with Chelating Agents. II. Fe_2O_3-low MW and Polymeric Hydroxamic Acid Species," *J. Colloid Interface Sci.* 75:199 (1980).

F-14. Burger, J. R. "Froth Flotation Developments," *Eng. Min. J.* 184:67–75 (1983).

F-15. Valdes-Krieg, E. et al. "Separation of Cations by Foam and Bubble Fractionation," *Sep. Purif. Methods* 6:221–285 (1977).

F-16. Jones, A. D. and A. C. Hall. "Removal of Metal Ions from Aqueous Solution by Dissolved-Air Flotation," *Filtr. Sep.* 18:386–390 (1981).

F-17. McDonald, C. and A. Suleiman. "Ion Flotation of Copper Using Ethyl (Hexadecyldimethyl) Ammonium Bromide," *Sep. Sci. Technol.* 14:219 (1979).

F-18. Yamada, K. et al. "Application of Monoalkyl Phosphates to Ion Flotation," *J. Chem. Soc. Jpn., Chem. Ind. Chem.* 10: 1900–1906 (1972).

F-19. Yamada, K., et al. "Application of N-Alkylethylendiamines to ion flotation," *J. Chem. Soc. Jpn., Chem. Ind. Chem.* 4:575–580 (1977).

F-20. Fuerstenau, D. W. "Fine Particle Flotation," in *Fine Particles Processing*, Vol. 1, P. Somasundaran, Ed. (New York: AIME, 1980), chap. 35.

F-21. Sebba, F. *J. Colloid. Sci.* 35(4):643 (1971).

F-22. Sebba, F. U.S. Patent 3900420 (1971).

F-23. Melville, J. B. and E. Matijevic. "Microbubbles: Generation and Interaction with Colloid Particles in Foams," in *Proceedings of the Symposium Society of Chemical Indus-*

try, Colloid and Surface Chemistry, R. J. Akers, Ed. (New York: Academic Press, 1976).

F-24. Ciriello, S. et al. "Removal of Heavy Metals from Aqueous Solutions Using Microgas Dispersions," *Sep. Sci. Technol.* 17:521 (1982).

F-25. Auten, W. L. and F. Sebba. "The Use of Colloidal Gas Aphrons (CGAS) for Removal of Slimes from Water by Floc Flotation," *Solid-Liquid Separation*, J. Gregory, Ed. (London and Chichester: Ellis Horwood, Ltd., 1984).

F-26. Allen, W. D. et al. "Adsorbing Colloid Flotation of Cu(II) with a Chelating Surfactant," *Sep. Sci. Technol.* 14:796 (1979).

F-27. Okamoto, Y. and E. J. Chou. "Chelation Effects of Surfactant in Foam Separation: Removal of Cadmium and Copper Ions from Aqueous Solution," *Sep. Sci. Technol.* 10:741 (1975).

F-28. Chou, E. J. and Y. Okamoto. "Concentration Effects on Separation Selectivity in Foam Fractionation," *Sep. Sci.* 13:439 (1978).

F-29. Chou, E. J. and Y. Okamoto. "Foam Separation of Mercury Ion with Chelating Surfactant—the Selectivity of the Removal of Cd and Hg Ions with 4 Dodecyl-diethyene Triamine," *Sep. Sci.* 10:741 (1975).

F-30. Huang, S. D. and D. J. Wilson. "Foam Separation of Hg(II) and Cd(II) from Aqueous Systems," *Sep. Sci.* 11:215 (1976).

F-31. Chatman, T. E., et al. "Constant Surface Charge Model in Floc Foam Flotation. The Flotation of Cu(II)," *Sep. Sci.* 11:141 (1979).

F-32. Barnes, J. C., et al. "Floc Foam Flotation of Ni, Cr, Co and Mn. Interaction in Surface Adsorption," *Sep. Sci. Technol.* 14:777 (1979).

F-33. Wilson, D. J. and E. L. Thackston. "Foam Flotation Treatment of Industrial Wastewaters: Laboratory and Pilot Scale." EPA-600/2-80-138 (1980).

F-34. Baarson, R. E. and C. L. Ray. "Precipitate Flotation—a New Metal Extraction and Concentration Technique," in *Unit Processes in Hydrometallurgy*, M. E. Wadsworth and T. T. Davis, Eds. (New York: Gordon & Breach, Science Publishers, Inc., 1964).

F-35. Beitelshees, C. P. et al. "Precipitate Flotation for Removal of Cu from Dilute Aqueous Solution," in *Recent Developments in Separation Science*, Vol. 5, N.N. Li, Ed. (CRC Press, Inc., Boca Raton, FL, 1979), chap. 4.

F-36. Mukal, S. et al. "Study on the Removal of Heavy Metal Ions in Waste Water by the Flotation Method," in *Flotation*—A.M. Gaudin Memorial Vol. 1 (New York: AIME, 1976), chap. 6.

F-37. Grieves, R. B. et al. "Ion Colloid and Precipitate Flotation of Inorganic Anions," in *Flotation*—A.M. Gaudin Memorial Vol. 1 (New York: AIME, 1976), chap. 11.

F-38. Perez, J. W. and A. F. Aplan. "Ion and Precipitate Flotation of Metal Ions from Solution," *AIChE Symp. Ser.* 71(150):34 (1975).

F-39. Mumallah, N. M. and D. J. Wilson. "Precipitate Flotation Studies with Monolauryl Phosphate and Monolauryl Dithiocarbamate," *Sep. Sci.* 16:213 (1980).

F-40. Kim, Y. S. and H. Zeitlin. "Separation of Zn and Cu from Seawater by Adsorption Colloid Flotation," *Sep. Sci. Technol.* 7:1 (1972).

F-41. DeCarlo, E. H. et al. "Recovery of Metals from Sulfated Deep-Sea Ferromanganese Nodules with Organic Precipitating Reagents and Adsorptive Bubble Techniques," *Sep. Sci. Technol.* 17:1023–1044 (1982–83).

F-42. Matsuzaki, C. and H. Zeitlin. "The Separation of Collectors Used as Coprecipitants of Trace Elements in Seawater by Adsorbing Colloid Flotation," *Sep. Sci. Technol.* 8:185 (1973).

F-43. Bleasdell, B. et al. "Separation of Metals from Sulfated Deep-Sea Ferromanganese Nodules with Organic Precipitating Reagents and Adsorptive Bubble Techniques," *Sep. Sci Technol.* 17:1635 (1982–83).

F-44. Sedgewick, P. et al. "Separation of Metals from Treated Deep-Sea Ferromanganese Nodules by Adsorptive Bubble Techniques Using Salicylaldoxime and Na Dithiocarbamate as Organic Precipitants," *Sep. Sci. Technol.* 19:183 (1984).

F-45. Chernykh, S. I. "Flotation Apparatus for Waste Water Purification," *Tsvetn. Met.* (11):101–104 (1978).

F-46. Formanek, J. and H. Holeckova. "Ion Flotation and its Application in Industry," *Freiberg. Forschungsh. A.* A593:111–123 (1978).

F-47. Dimitrova, M. and S. Stoev. "Description of Technology and Apparatus for Extraction of Copper from Waste Water by Sediment Flotation," *God. Vissh. Minno-Geol. Inst., Sofia* Vol. Date 1975–76 22(Pt. 4):35–44 (1978).

F-48. Nazarova, G. N., L. V. Kostina, and L. N. Dorokina. "Use of Electrolytic Flotation for Purifying Waste Waters from Concentration and Metallurgical Plants with Simultaneous Complete Extraction of Valuable Components," *Mater. Nauchno-Tekh. Soveshch. "Kompleksn. Ispol'z. Syr'evykh Resur. Predpr. Tesetn. Metal."*, Meeting Date 1974. A. Adibekyan, Ed. (1977), pp. 299–325.

F-49. Sviridov, V. V. and A. I. Gomzikov. "Flotation Separation of Transition Metals from Ethylenediamine Solutions," *Kompleksn. Ispol'z. Miner. Syr'ya* (3):58–62 (1980).

F-50. Agency of Industrial Sciences and Technology. "Selective Separation of Aqueous Metal Ions in High Concentration." Japanese Patent No. 55158237 (1980).

F-51. Jude, E. "Removal of Uranium, Copper, Zinc and Molybdenum Ions from Wastewaters." Romanian Patent No. 71647 (1980).

F-52. Nippon Mining Co., Ltd. "Metal Recovery from Industrial Wastewater." Japanese Patent No. 55037955 (1980).

F-53. Jones, A. D. and A. C. Hall. "Removal of Metal Ions from Aqueous Solution by Dissolved-Air Flotation," *Filtr. Sep.* 18(5):386–388, 390 (1981).

F-54. Schulz, R. "Metal Recovery from Complex Ores and Secondary Raw Materials." Federal Republic of Germany Patent No. 3025740 (1982).

F-55. Letowski, F., S. Michalak, G. Sokalska, J. Drzymala, and J. Mordalski. "Utilizing Wastes from Acid Lixiviation of Copper Concentrates." Polish Patent No. 110494 (1981).

F-56. Eliseev, N. I., G. P. Kharitidi, V. P. Yuferov, and G. V. Skopov. "Flotation-Metallurgical Treatment of Carbon Containing Dusts," *Kompleksn. Ispol'z. Miner. Syr'ya* (4):58–61 (1982).

F-57. Kalenga, N. and C. Ek. "Recovery of Copper Contained in Luilu Leachating Wastes [Zaire]," *Maadini* 25:46–48 (1981).

F-58. Stocica, L. and A. Duca. "Recovery of Copper from Wastewater." Romanian Patent No. 75393 (1980).

F-59. Miskovic, D., E. Karlovic, and B. Dalmacija. "The Investigation of Application of Dissolved Air Precipitate Flotation in the Absence of Collector and Frother for the Purification of Wastewater Containing Metal Ions," *Stud. Environ. Sci.* 23 (*Chem. Prot. Environ.*): 245–252 (1984).

F-60. Brooks, C. S. "Metal Recovery from Waste Sludges," *Proceedings of the 39th Industrial Waste Conference* (Boston: Butterworth Publishers, Inc. [Ann Arbor Science Book], 1985), pp. 529–535.

F-61. Collee, R., G. Monfort, and G. Windels. "Beneficiation of Cobalt Ores – Experimental Study of a Deposit," *Ann. Mines Belg.* (3–4):105–131 (1985).

F-62. Boteva, A. and V. Kovacheva. "Selective Flotation of Sulfide Precipitates," *God. Vissh. Minno-Geol. Inst., Sofiya*, Volume Date 1983–1984, 30(4):227–236 (1984).

F-63. Cherkasov, A. E., N. N. Voronin, and G. N. Dobrokhotov. "Kinetics of Ionic Flotation and the Mixed [Adsorptive-Adhesive] Mechanism of the Process," *Izv. Vyssh. Uchebn. Zaved. Tsvetn. Metall.* (2):3–8 (1987).

F-64. Kunicki-Goldfinger, W., A. Lejczak, and M. Ostrowski. "Leaching of Heavy Metals from Post-Flotation Wastes and/or Low-Grade Ores." Polish Patent No. 130762 (1985).

F-65. Molnar, L. "Recovery of Copper from Slag Dump by Cooling Technology," *Zesz. Nauk. Akad. Gorn.-Hutn. Stanislawa Staszica. Metal. Odlew.* (109):433–440 (1987).

F-66. Demidov, V. D., A. M. Berestovoi, N. N. Voronin, V. I. Tolstunov, A. E. Cherkasov, D. D. Uspenskii, V. N. Kravchenko, Y. I. Mishchenko, P. S. Kislitsyn, and T. D. Rogachkova. "Recovery of Copper from Solutions by Flotation." U.S.S.R. Patent No. 1477758 (1989).

F-67. Sorensen, J. L., M. D. Yarbro, and C. A. Glocker. "Method for Removal of Organic Solvents from Aqueous Process Streams." U.S. Patent No. 8906161 (1989).

F-68. Stoica, L. and L. Magyar. "Flotation Separation of Copper(II) from Solutions Diluted with Romegal CM," *Rev. Chim. (Bucharest)* 39(12):1103–1107 (1988).

F-69. Atkinson, R. J., A. L. Hannaford, L. Harris, and T. P. Philip. "Using Smelter Slag in Mine Backfill," *Mines Mag.* 160(8):118–123 (1989).

F-70. Rubin, A. J., J. D. Johnson, and J. C. Lamb, III. "Comparison of Variables in Ion and Precipitate Flotation," *Ind. Eng. Chem. Proc. Des. Dev.* 5, 368 (1966).

F-71. Somasundaran, P. and R. B. Graves, Eds. "Advances in Interfacial Phenomena of Particulate/Solution/Gas Systems: Applications of Flotation Research," *AIChE Symp.* 71, (150):34 (1975).

F-72. McIntyre, G. T., J. J. Rodreguez, E. L. Thackston, and D. J. Wilson. "The Removal of Mixtures of Metals by an Adsorbing Colloid Foam Flotation Pilot Plant," *Sep. Sci. Technol.* 17(1):683 (1982).

F-73. Kawalec-Pietrenko, B. and A. Selectel. "Investigations of Kinetics of Removal of Trivalent Cr Salts from Aqueous Solutions using Ion and Precipitate Flotation," *Sep. Sci. Technol.* 19(13–15):1025 (1985).

F-74. Rubin, A. J. and J. D. Johnson. "Effect of pH on Ion and Precipitate Flotation Systems," *An. Chem.* 39:298 (1967).

F-75. Huang, S. D., M. K. Huang, J. Y. Gua, T. P. Wu, and J. Y. Kung. "Simultaneous Removal of Heavy Metal Ions from Wastewater by Foam Separation Techniques," *Sep. Sci. Technol.* 23(4 & 5):489 (1988).

F-76. Fuerstenau, A. W. and R. Herrera-Urbina. "Mineral Separation by Froth Flotation," in *Surfactant-Based Separation Processes*, J. F. Scamehorn and J. H. Harwell, Eds. (New York: Marcel Dekker, Inc., 1989), chap. 11.

F-77. Sebba, F. "Novel Separations Using Aphrons," in *Surfactant-Based Separation Processes*, J. F. Scamehorn and J. H. Harwell, Eds. (New York: Marcel Dekker Inc., 1989), chap. 4.

6.3 MAGNETIC SEPARATIONS

High-gradient magnetic separation technology permits separation and concentration of soluble metals and colloidal systems and solid particle with sizes to about 1 μm. Application can be made to ferromagnetic metals such as iron, cobalt, and nickel and paramagnetic metals such as chromium, manganese, molybdenum, titanium, vanadium, tungsten, tantalum, uranium, rare earths, and precious metals.

Magnetic separations have been applied to a wide variety of waste systems such as scrap metals, dusts, electrical and electronic scrap, battery components, grinding waste, municipal wastes, automotive scrap, mine wastes, various metallic sludges, and metals in solution (Table 6.3). The magnetic systems used have consisted of eddy current separators[M-5, M-7, M-14, M-17] and magneto hydrostatic[M-8, M-12, M-15] separators, and in some instances high temperature heating (500 to 1000°C) has preceded magnetic separation.[M-11, M-20, M-22] Magnetic separations have also been combined with gravity segregation[M-3, M-16] and electrolysis.[M-4, M-5]

Table 6.3 Metal Recovery by Magnetic Separation

Waste system	Metals	Special features	Metal separation efficiency %	Ref.
Wastewater	Fe, Cu	Electrolysis with steel cathode and magnetic anode	80–90 Cu/Fe	M-4
Electrochemical reactor	Cu	Electrolysis with eddy promoters		M-5
Electronic	Cu, Sn, Ni, Pb, W, Mo	Magnetic-hydrostatic separation		M-8
Wastewater	Cu, Cr, Cd, Hg, Fe	Magnetic separation metal sulfides		M-9
Mine waste	Co, Cr, Ni	Applicable to kyanite and chromite—combination gravity/magnetic separation	Co recovery	M-10
Municipal/auto scrap	Cu, Pb	Eddy current separators	98 Cu, 94 Pb	M-7
Steel dust/sludge	Ni	Hot melt process—mostly recovers Ni using magnetic separation		M-11
Waste cables	Cu	Magneto-hydrostatic separation	99.6 Cu	M-12

Table 6.3 continued

Waste system	Metals	Special features	Metal separation efficiency %	Ref.
Electrical scrap	Al, Cu, Pb	Eddy current separator	<90 as Al alloy	M-14
Electronic	Al, Cu	Combination gravity segregation/ magnetic separation		M-16
Cable scrap	Cu, Pb	Eddy current separator for Cu/Pb		M-17
Cu dust	Cu, Pb, Cd, As, Co, In Bi	H_2SO_4 leach followed by magnetic separation	50-60 Zn, 60 Cd, 80 Pb/As/Sn	M-19
Metal scrap	Al, Cu, Fe,	Combination electrodynamic and magnetic separation	95.3 nonmagnetic Cu/Al and magnetic Cu/Al/Fe	M-21
Battery	Fe, Mn, Zn, Cu, Ag	Involves heating 500 to 1100° and magnetic separation of Fe/Cu/Ag/Mn/Zn		M-22
Shredded scrap	Cu, bronze, Sn	Magnetic + float sink in heavy medium	50-80 Cu + Zn, 78 Al	M-24

Ferrite coprecipitation,[M-1, M-2, M-3] a special precipitation process used in combination with magnetic separation, is applicable to removal of soluble metals from waste effluents. In this process, advantage is taken of the ease of coprecipitation of various metals, such as cadmium, copper, chromium, nickel, mercury, tin, manganese, zinc, and lead, with ferric hydroxide. An advantage of this process is that when a ferromagnetic precipitate is obtained, high-gradient magnetic separation technology can be used to facilitate metal recovery.

References

M-1. Okamato. S. "Iron Hydroxides as Magnetic Scavengers," *IEEE Trans. Magn.* 10:923 (1975).

M-2. Okuda, T. et al. "Removal of Heavy Metals from Wastewater by Ferrite Co-Precipitation," *Filtration Separation* 12:472 (1975).

M-3. Shimoiizaka, J., K. Nakatsuka, T. Fujita, and A. Kounosu. "Sink-Float Separators Using Permanent Magnets and Water Based Magnetic Fluid," *IEEE Trans. Magn.* MAG-16(2):368–371 (1980).

M-4. Fujishiro, M. "Investigation of Separation and Recovery of Heavy Metals from Waste Water by Electrolytic Method," *Kankyo Kagaku Kenkyu Hokoku* (*Chiba Daigaku*). Volume Date 1975–1976, 3:34–35 (1978).

M-5. Storck, A. and D. Hutin. "Improvement of Copper Recovery in Electrochemical Reactors using Turbulence Promoters," *Electrochim Acta* 26(1):117–125 (1981).

M-6. Ambrose, F. and B. W. Dunning, Jr. "Mechanical Processing of Electronic Scrap to Recover Precious-Metal-Bearing Concentrates," in *Precious Metals* [Proc. Int. Precious Met. Inst. Conf.], 4th Meeting Date 1980 R. McGachie, O. Richard, and A. G. Bradley, Eds. (Willowdale, Ontario, Pergamon, Canada, 1981), p. 67.

M-7. Scholemann, E. "Eddy-Current Techniques for Segregating Nonferrous Metals from Waste," *Conserv. Recycl.* 5(2–3), 149–162 (1982).

M-8. Kravchenko, N. D., Y. M. Dubinskii, and V. I. Krichevskii. "Processing of Electronic Scrap," *Tsvetn. Met.* (3):86–87 (1983).

M-9. Nakayama, M. "Heavy Metal Recovery from Wastewater as Sulfide." Japanese Patent No. 59069191 (1984).

M-10. Mani, K. S. and D. Subrahmanyam. "Utilization of Low-Grade/Off-Grade Ores and Mine Wastes," *Proc. Indian Natl. Sci. Acad., Part A.* 50(5):509–522 (1984).

M-11. Tanigawa, K. and T. Oshiumi. "Recovery of Metals from Steelmaking Dust and Sludge." Japanese Patent No. 60159133 (1985).

M-12. Kravchenko, N. D., N. A. Bondarev, and A. V. Dubinin. "Effect of the Temperature of the Ferromagnetic Liquid on the indicies of Magnetohydrostatic Separation," *Tsvetn. Met. (Moscow)* (12):73–75 (1986).

M-13. Kravechenko, N. D. "Magnetohydrostatic Separation of Nonferrous Metal Scrap into Fractions of Different Density," *Freibert. Forschungsh. A.* A 745:125–128 (1986).

M-14. Bredikhin, V. N., O. M. Cherepnin, and A. I. Shevelev. "Eddy Current Classification of Nonferrous Metal Scrap," *Freibert. Forschungsh. A.* A 745:118–124 (1986).

M-15. Hrabak, V., P. Vejnar, V. Horak, F. Stybal, Z. Kenclova, and J. Malina. "Recovery of Precious Metals from Waste Materials," Czechoslovakian Patent No. 232583 (1987).

M-16. Hurop, G., G. Schubert, and F. Hartwig. "Processing of Scrap from Electrical Engineering and Electronics," *Freibert. Forschungsh. A.* A 746:103–114 (1987).

M-17. Dalmijn, W. L. and H. J. L. Van der Valk. "Separation of Nonferrous Metals from Mixtures Using Permanent Magnets in a Vertical Eddy-Current Separator," *Neue Huette* 32(7):273–279 (1987).

M-18. Sikora, B. and J. Steindor. "Recovery of Nickel and Cobalt from Grinding Suspensions," *Freibert. Forschungsh. A.* A 746:56–57 (1987).

M-19. Lastra-Quintero, R., N. Rowlands, S. R. Rao, and J. A. Finch. "Characterization and Separation of a Copper Smelter Dust Residue," *Can. Metall. Q.* 26(2):85–90 (1987).

M-20. Heng, R., W. Koch, and H. Pietsch. "Processing of Spent Miniature Batteries." European Patent No. 244901 (1987).

M-21. Barskii, L. A. and I. M. Bondar. "Electrodynamic Separation of Nonmagnetic Electrically Conducting and Ferromagnetic Particles," *Tsvetn. Met.* (*Moscow*) (6):87–89 (1988).

M-22. Heng, R., W. Koch, and H. Pietsch. "Recycling of Spent Batteries." Federal Republic of Germany Patent No. 3709967 (1988).

M-23. Morency, M. "Production of Elements and Compounds by Deserpentinization of Ultramafic Rock." U.S. Patent No. 4798717 (1989).

M-24. Vejnar, P. and V. Hrabak. "Recovery of Nonferrous Metals from Comminuted Scrap by Sink-Float Separation," *Freiberg. Forschungsh. A.* A 746:85–2, (1987).

M-25. Sikora, B. and L. Szymocha. "Recovery of Alloying Elements from Grinding Swarf in Production of Aluminum-Nickel and Aluminum-Nickel-Cobalt-Type Magnetic Alloys," *Hutnik* 51(10):376–377 (1984).

M-26. Kelland, D. "High Gradient Magnetic Separation Applied to Mineral Benefication," *IEEE Trans. Magn.* 9(3):307 (1973).

M-27. Obertenffer, J. "Magnetic Separation: A Review of Principles, Devices, and Applications," *IEEE Trans. Magn.* 10(2):223 (1974).

M-28. Obertenffer, J., I. Wechsler, P. G. Marston, and M. J. McNallon. "High Gradient Magnetic Filtration of Steel MOH Process and Waste Waters," *IEEE Trans. Magn.* 11(5):1591 (1975).

M-29. DeLatouri, C. and H. Kolm. "Magnetic Separation in Water Pollution Control II," *IEEE Trans. Magn.* 11(5):1570 (1975).

6.4 PYROMETALLURGICAL SEPARATIONS

Inasmuch as this technology review is primarily concerned with hydrometallurgical processes, pyrometallurgical processes are considered only inasmuch as they provide preparative stages in the hydrometallurgical separation processes examined. The roles played by pyrometallurgy are quite varied, consisting of oxidation, reduction, thermal decomposition, volatilization of waste components, and facilitation of reactions such as sulfurization or chlorination, and achieving separations of molten components as alloys. Some examples for a wide range of waste systems are summarized in Table 6.4.

References

H-1. Stephenson, J. B., A. A. Cochran, and P. G. Barnard. "Recovery of Zinc Oxide from Galvanizing Wastes," *Proceedings of the Sixth Mineral Waste Utilization Symposium.* U.S. Bureau of Mines and IIT Research Institute, May 23 (1978), p. 320.

H-2. Witzke, L. and W. Mueller. "Treatment of Nonferrous Metal Hydroxide Slime Wastes." German Patent No. 2743812 (1977).

H-3. Maass, W., H. Kirschner, G. Loeffler, and H. Gerhard. "Treatment of Fine-Granular Oxidized or Metallic Secondary Copper Materials." German Democratic Republic Patent No. 134120 (1979).

H-4. Brun, D., G. Chrysostome, A. Feugier, and B. Sale. "Process for Recovering Metal Elements from Carbonaceous Products." U.K. Patent No. 2026458 (1978).

Table 6.4 Metal Separation by Pyrometallurgical Processes

Waste system	Metals	Pyrometallurgy temp./atmos.	Subsequent processing	Metal separation efficiency %	Ref.
Hydroxide slime	Cr, Cu, Ni, Zn	400-800°C/air + hydroxide and carbonate	H_2SO_4 leach/liq.-liq extr		H-2
Cu wastes	Cu, Zn	1050-1150°C/air/C/SiO_2	Zn sublimation/Cu melt at 1450°C	Cu 92 pure	H-3
Fuel soot	Ni, V	Oxid. 870°C/$CaCO_3$/SiO_2		Ni/V alloy	H-4
Scrap metal	Co, Fe, Ni	Melt with chromite/magnesite	Separation alloy and slag	95-99 Ni; 50-90 Co	H-5
Zn waste clinker	Ag, Au, Cu, Zn	Chloride/sublimation roast	Powder sizing	29 Cu; 3.5 Pb; 18 Zn; + 14.2 g Au/ton	H-7
Waste slag	Co, Ni	Electric depletion furnace melt	Slag flotation	13-26 Co; 28-77 Ni	H-8
Spent petroleum catalyst	Ni	Hydrometallurgical roast to 600°C Pyrometallurgical roast to 600°C	H_2SO_4 leach	Ni recovery to 96	H-9
Superalloy scrap	Co, Cr, Mo, Ni, Nb, Ta	Induction furnace melt	Leach and flotation	96 Co; 99 Cr; 92 Mo; 99 Ni	H-10
Waste slag	Cu	Roast to 800°C/$CaCO_3$/SiO_2/C			H-11
Catalyst and electrical scrap	Ag, Cu	Air roast to 1400°C with flux		91 Ag; 77 Cu	H-12
Plating hydroxide waste	Ni	1490°C melt with reduction agent		91 Ni in product	H-13
Spent catalyst	Al, Co, Mo, Ni, V	Heat to 560°C with alkali sulfates		97 Co; 92 Mo	H-14

Table 6.4 continued

Waste system	Metals	Pyrometallurgy temp./atmos.	Subsequent processing	Metal separation efficiency %	Ref.
Scrap alloy	Cr	Melt to 1450°C with O_2	HCl leach/electrolysis	Cr_2O_3 H_2O product	H-18
Cu electrolyte	As, Bi, Cu, Sb	H_2 reduction at 180°C	As separation	90 of Cu	H-19
Electrolysis sludge	Ag, Cu, Ni, Se, Fe	Oxid. roast to 700°C	99 Se separation/leach	Cu recovery	H-21
Automotive scrap	Cr, Cu, Ni	Oxid. roast to 1700°F/red. to 1900°F	Sizing/magnetic and flotation separation	50 Cu; 54 to 96 Cr; 47 to 98 Ni	H-22
Steel scrap	Al, Cu, Pb, Sn	Melt to 800°C/coke/lime	Sizing	Cu concentration in slag	H-23
Steel dust and sludge	Fe, Ni, Cr	Melt with chloride/coke	Size and magnetic separation	Cr, Fe, Ni, concentration in alloy	H-26
Battery waste	Ag, Hg, Ni, Zn	Zn alloy anode heat to 650°C	Hg vaporization	Ni steel scrap and Zn	H-28
Magnetic alloy grinding sludge	Cu, Co, Fe, Ni	950°C H_2 red. to 1200°C	Solvent extraction/oxid. to separate Al_2O_3 slag	Co/Fe/Ni 95 + Cu 90	H-30
Cu waste	Cu	Chloride/C/roast to 750°C	Flotation of slag	92 of Cu to coke	H-33
Varied industrial waste	Co, Cu, Fe, Ni, Pb, Sn, Zn	Melt to 1450°C/C/sulfates	Sublimation Pb, Sn, Zn + size	99 Ni; 84-95 Co; 90-96 Cu	H-38
Circuit board waste	Cu	Heat to 800°C/red. waste gas		95 Cu in fly ash	H-40
Cu waste slag	Cu, Fe	Chloride roast to 800°C	Flotation of slag	97 of Cu in slag	H-42
Metal wastes	Co, Cr, Mn, Ni	CO atmos. 140°C	Thermal decomposition of metal carbonyls	Conc Cu, Cr, Mn, Ni alloy	H-45

H-5. Astaf'ev, A. F. and N. I. Solovov. "Melting of Cobalt-Containing Secondary Raw Material for Nickel Black," *Tsvetn. Met.* (7):70–71 (1980).

H-6. Kurzejs, A. and S. Wojcik. "Production of Silver from Raw Materials and Recycled Materials," *Wiad. Hutn.* 36(5):153–157 (1980).

H-7. Maiskii, O. V., V. G. Voronin, and V. I. Petunin. "Pilot-Plant Tests of a Method of Clinker Processing by Chloride-Sublimation Roasting," *Tsvetn. Met.* (11):39–43 (1980).

H-8. Piotrovskii, V. K., Y. L. Serebryanyi, V. L. Pichkur, and G. D. Petrova. "Forms in Which Nickel and Cobalt Occur in Waste Slags of Depletion Melting," *Tsvetn. Met.* (2):24–26 (1981).

H-9. Zambrano, A. R. "Recovery of Nickel from Wastes." German Patent No. 3000040 (1980).

H-10. DeBarbadillo, J. J., J. K. Pargeter, and H. V. Makar. "Process for Recovering Chromium and Other Metals from Superalloy Scrap," Rep. Invest. – U.S. Bureau of Mines, RI 8570 (1981).

H-11. Bunatyan, E. G., S. A. Bakhchisaraiteseva, I. E. Ovanesova, O. N. Shakhbazyan, T. K. Osipova, and A. A. Davtyan. "Effect of Ferrite Formation in Calcines on Losses of Copper with Waste Slags," *Promst. Arm.* (2):40–42 (1982).

H-12. Van Hecke, M. C. F. and L. M. Fontainas. "Extraction of Nonferrous Metals from Residues Containing Iron." European Patent No. 53406 (1982).

H-13. Eichberger, E. and G. Lazar. "Recovery of Heavy Metals." European Patent 60826 (1982).

H-14. Rastas, J. K., K. J. Karpale, and H. Tiitinen. "Recovery of Metal Values from Spent Catalysts Used in Extracting Sulfur from Crude Petroleum." Belgian Patent No. 894678 (1983).

H-15. Lach, E. and Z. Myczkowski. "Pressureless Extraction of Materials Difficult to Oxidize," *Pol. Tech. Rev.* (5–6):18–19 (1982).

H-16. Voronin, N. M., L. M. Soifer, V. M. Moshenskii, and L. I. Netimenko. "Physicochemical Properties of Molten Waste Slags After Extensive Copper Extraction," *Tsvetn. Met.* (4):24–25 (1983).

H-17. Mattheaus, R., D. Koennecke, S. Mehwald, and J. Dauderstedt. "Apparatus for Remelting and Refining Low-Grade Copper Scrap." German Democratic Republic Patent No. 200528 (1983).

H-18. Pu, L. and Z. Yan. "Recovery of Chromium from Scrap Alloy Steel," *Youse Jinshu* 34(5):1–5 (1982).

H-19. Togashi, R. and T. Nagai. "Hydrogen Reduction of Spent Copper Electrolyte," *Hydrometallurgy* 11(2):149–163 (1983).

H-20. Daido Steel Co., Ltd. "Metal Recovery from Incineration Ash and Sludges." Japanese Patent 58037371 (1983).

H-21. Ryzhov, A. G., V. D. Chegodaev, and Z. A. Eiteneer. "Oxidation-Roasting of a Sludge from Processing of Copper-Nickel Raw Material," *Tsvetn. Met.* (5):15–17 (1984).

H-22. Herter, C. J. "Low Residual Alloy Steel Charge from Scrap Metal." U.S. Patent No. 4517016 (1985).

H-23. Eckstein, H., W. Viehl, R. Wagenmann, H. Knoll, O. Siebenhuener, G. Hocke, J. Simon, W. Albrecht, G. Peter et al. "Treatment of Copper-Containing Iron Scrap." German Democratic Republic Patent 215586 (1984).

H-24. Koch, W., W. Tuerke, and H. Pietsch. "Processing of Small Batteries." Federal Republic of Germany Patent 3402196 (1985).

H-25. Tanigawa, K. and T. Oshiumi. "Recovery of Metals from Steelmaking Dust and Sludge." Japanese Patent No. 60159130 (1985).

H-26. Tanigawa, K. and T. Oshiumi. "Recovery of Metals from Steelmaking Dust and Sludge." Japanese Patent No. 60159128 (1985).

H-27. Tanigawa, K. and T. Oshiumi. "Recovery of Metals from Steelmaking Dust and Sludge." Japanese Patent No. 60159129 (1985).

H-28. Uenishi, T. and J. Inoue. "Material Recovery from Spent Silver Oxide Batteries." Japanese Patent No. 60036827 (1985).

H-29. Tomasek, K. and J. Schmiedl. "Use of Chloride Metallurgy in the Processing of Secondary Raw Materials," *Freiberg. Forschungsh. B.* B 251:16–24 (1985).

H-30. Holman, J. L. and L. A. Neumeier. "Reclamation of Metals from Magnet Alloy Grinding Sludge," in *Recycle Second. Recovery Met., Proc Int. Symp.* P. R. Taylor, H. Y. Sohn, and N. Jarrett, Eds. (Warrendale, PA: Metallurgical Society, 1985), pp. 327–336.

H-31. Abe, T. "Purification of Crude Gold Recovered from Copper Electrolysis Sludge." Japanese Patent No. 61037932 (1986).

H-32. Holman, J. L. and L. A. Neumeier. "Experimental Nickel-Cobalt Recovery from Melt-Refined Superalloy Scrap Anodes," Rep. Invest. – U.S. Bureau of Mines , RI 9034 (1986).

H-33. Molnar, L., E. Kassayova, S. Cempa, K. Tomasek, J. Simko, L. Wedinger, and L. Miklos. "Method of Copper Extraction from Copper Works Waste." Czechoslovakian Patent No. 234898 (1987).

H-34. Morency, M. "Production of Elements and Compounds by Deserpentinization of Ultramafic Rock." U.S. Patent No. 8701731 (1987).

H-35. Riveros, U. G., G. Luraschi, and A. Antonio. "Recovery of Copper from Reverberatory Waste Slags in an Electric Arc Furnace," *Minerales* 42(179):53–60 (1987).

H-36. Cohen, J., M. Demange, J. M. Lamerant, and L. Septier. "Process for Recovering Aluminum and Lithium from Scrap Metal." European Patent No. 250342 (1987).

H-37. Hein, K. "Hydrometallurgical Treatment of Copper-Containing Secondary Raw Materials." *Freiberg. Forschungsh. B.* B 260:41–51 (1987).

H-38. Rudorf, M., L. Mueller, W. Dittrich, and W. Goetzelt. "Processing of Nickel-, Cobalt-, and Copper-Containing Scrap and Industrial Wastes." German Democratic Republic Patent No. 250137 (1987).

H-39. Sukla, L. B., S. C. Panda, and P. K. Jena. "A Process for Extraction of Cobalt, Nickel, and Copper Metal Values from Copper Converter Slags and Similar Industrial Wastes." Indian Patent No. 160520 (1987).

H-40. Tschentke, J., R. Ohlendorf, W. Albrecht, G. Landwehr, I. Wetzel, M. Kunis, and R. Herzschuch. "Treatment of Wastes from Copper-Coated Printed-Circuit Boards." German Democratic Republic Patent No. 258141 (1988).

H-41. Ladd, J. A. and M. J. Miller. "Recovery of Chromium and Other Metal Values from Superalloy Scrap." U.S. Patent No. 4798708 (1989).

H-42. Molnar, L. and S. Cempa. "Application of the Segregation Process to Roasting of Waste Copper Slags," Zb. Ved. Pr. Vys. Sk. Tech. Kosiciach (2):237–246 (1988).

H-43. Oden, L. L. and G. W. Elger. "Removal of Copper from Molten Ferrous Scrap: Results of Laboratory Investigations." U.S. Bureau of Mines, RI 9139 (1987).

H-44. Hori, M., K. Yaku, and K. Nagdya. "Treatment of the Sludge Containing Cr and Ca by Heating with Silica," Environ. Sci. Technol. 12(13):1431 (1978).

H-45. Visnapu, A., and L.C. George. "Recovery of critical metals by carbonyl processing," U.S. Bureau of Mines RI 9087 (1987).

6.5 SOLVENT PARTITION

Solvent partition is a technique currently under experimental evaluation that has promising potential for recovery of insoluble metal hydroxides in the colloidal particle size range. Insoluble metal hydroxides in the small particle size range can be transferred from an aqueous phase to a hydrocarbon phase by surfactant adsorption to modify wettability.

This process has been designated as *solvent partition* and is analogous to adsorbing colloid flotation where transfer is effected by alteration from hydrophilic to oleophilic wettability by adsorption of appropriate surfactants (C_{12} to C_{18} carboxylic acids).[SP-1]

Bench scale performance tests conducted with single and multicomponent aqueous metal salt solutions and metal finishing industry waste sludges have demonstrated efficient transfers to a hydrocarbon phase employing long-chain carboxylic acids (lauric and oleic acids) in the pH range 6 to 9 for solids loadings of several weight percent and with surfactant/solids loadings approximating saturation adsorption. Transfer is nonselective, but has a useful potential for efficient separation and dewatering of fine-particle sludges. Separation and dewatering of electrochemical machining colloidal metal oxides from brine can be achieved very efficiently by solvent partition if solids loading of the hydrocarbon phase does not exceed about 3 wt% solids. Solid loadings much above 3 wt% in the hydrocarbon phase produce excessive viscosity and thixotropic gels, occluding aqueous phase and frustrating dewatering of the separated solids. A possible application of solvent partition is the separation and dewatering of the colloidal metal hydroxide wastes of electrochemical milling.

References

SP-1. Brooks, C. S. "Adsorption Stabilization of Hydrocarbon Dispersons of Transition Metal Oxides," Paper No. 44, in *58th Colloid and Surface Science Symposium* (Pittsburgh: American Chemical Society, 1984).

7

Solubilization of Solid Wastes

Several alternatives are available for solubilizing the potentially wide range of waste solids ranging from scrap metal, appliance and hardware components, electrical and electronic scrap, mineral wastes, water purification sludges, flue dusts, fuel, and combustion ashes and residues. These solubilizing alternatives consist of reaction with acids, alkalies, chlorination, sulfidization, alloying, and formation of volatile compounds with carbon monoxide, and volatilization by heating for some metals and their compounds. Solubilization by certain biological reductions is also possible, as mentioned in Chapter 6.

Some examples of results obtained for these various solubilization routes are given in Table 7.1 and discussed below, starting with acid solubilization, which is probably the most commonly used alternative.

Additional examination is appropriate here concerning experience with metal recovery by acid leaching for a variety of solids with low metal contents, ranging from mining wastes and coal ash to municipal waste residues.

Acid solubilization is one of the most obvious initial preparation stages for conditioning metal waste sludges for application of the recovery processes discussed above for metals in solution. Information on acid solubilization of metal oxides in waste from extractive metallurgical processes,[SO-1 to SO-9, SO-12, SO-16, SO-23, SO-32, SO-38, SO-45] coal and petroleum ash,[SO-13, SO-14, SO-17, SO-35, SO-43] and sewage sludges[SO-15, SO-19, SO-31, SO-33] provides an indication of the efficiency of metal solubilization on waste

Table 7.1 Solubilization of Solid Wastes

Waste system	Metals	Acid leach	Alkali leach	Heat roast	Metal separation efficiency %	Ref.
Ni/Zn battery	Ni/Zn	H_2SO_4		To 90°C	>90 Ni	SO-11
Cu smelter slag	Cu, Zn	H_2SO_4			~85 Cu and Zn	SO-12
Coal ash	Al, Ba, Cd, Mg, Ce, Co, Cr, Cu, Mn, Ni, Nb, Mo, Ti	HCl, HNO_3, H_2SO_4		Reflux temperature	50 Al; 30 Ti; 60–80 heavy metals	SO-14
Municipal sewage sludge	P, Fe, Al, Mg, Mn, Ni, Zn, Cr, Cu, Pb	H_2SO_4			80 of P, Fe, Al, Zn, Mg, Mn, Ni	SO-15
Metal oxides/silicates	Li	HCl	CaO/MgO	Chloride roast to 800°C	90 of Li	SO-23
Grinding sludge	Co, Cu, Ni, Fe			Oxid. and H_2 roast to 1000°C	99 of Co, Cu, Ni, Fe	SO-24
Sulfide ores	Cu, Ni, Fe	H_2SO_4	$Fe_2(SO_4)/H_2SO_4$	To 70-100°C	99 Cu and Fe—no Cu, precious metals	SO-25
Pb waste	Cu, Pb		$NH_4OH/(NH_3)_2CO_3$		90 of Cu	SO-27
Fly ash	Cu, Zn	Waste acid	Alkali		Fertilizer residue (Cu, Zn, and acceptable)	SO-28
Plating waste	Cu		NH_4HCO_3		99 of Cu	SO-29

Waste	Metals				No.
Plating sludge	Cu, Ni, Zn		Ammonium liquor	Cu/Ni by solv. extr./Zn hydroxide carbonate	SO-30
Sewage sludge	Ag, Au, Cu	H_2SO_2/CN		Au/Ag cyanides recovered	SO-31
Waste oxides	Cu	Sulfides	Carbonates	Final leach with nitrilotriacetic acid	SO-32
Plating waste	Cr, Cu, Ni, Zn	Cl_2	NH_4OH/NaOH	Metal in solution in leachate	SO-34
Coal ash	Al + other metals	HCl/O_2		Mn, Ni, Zn	SO-35
Flue dust	As, Cd, Cu, Zn	H_2SO_4	NaHS + carbonate	90 of As, 95 of Cd, 94 of Cu, 93 Zn	SO-36
Cu waste	Cu, Re	H_2SO_4/H_2O_2		70 of Re + Cu as $Cu(OH)_2$	SO-41
Magnetic tape	Co, Ni		NaOH		SO-42
Petroleum coke	Ni, V	H_2SO_4	NH_3, $Ca(OH)_2$		SO-43

Table 7.1 continued

Waste system	Metals	Acid leach	Alkali leach	Heat roast	Metal separation efficiency %	Ref.
Spent catalyst	Co, Mo, Ni, V	$FeCl_3$	Na_2CO_3	Roast	90 of Cu/Ni, 97 of Mo	SO-44
Mn nodules	Cu, Ni		NH_4SCN		90 of Cu/Ni, 50–70 of Cu	SO-45
Spent catalyst	Ni			Carbonyl vapor	92–95 of Ni	SO-46
Blast furnace slag/dust	Al, Cr, Ni, Pb, Sn, Zn	Acid + O_2 + H_2O_2	KHS_2O_8		Major portion of metals recovered from leachates as hydroxides	SO-47
Coverter dust	Cu, Pb, Sn, Zn		$NH_4OH/(NH_4)_2CO_3$	Roast to 400°C	Most Zn as ZnO by-product and Cu, Pb, Zn as recycle cement	SO-26
Ore or smelter dust	Ag, Cu, Pb, Zn			O_2 roast to 80–150°C + $CaCl_2$	75–91 Cu, 71–82 Pb, 65–96 Zn, 67 Ag	SO-16
Coal ash		Acid	Lime	Roast		SO-17
Automotive catalyst	Pt, Pd, Rh	H_2SO_4		Roast with Cl_2 or SO_2	80–97 Pt, 80–96 Pd, 78–95 Rh	SO-18

sludges similar in some respects to those of the metal finishing industry. Uranium mining by-product wastes containing as little as 0.2 wt% vanadium have been successfully leached with sulfuric acid in preparation for solvent extraction recovery of the vanadium.[SO-3] Sulfuric acid leach provides promising recoveries of copper and zinc in copper smelter slag[SO-12] and converter dust.[SO-26]

Studies have been conducted regarding the potential of metal recovery of acid leachate from coal ash and petroleum.[SO-13, SO-14, SO-17, SO-35] Typical extraction efficiencies for metals of interest, such as cobalt, chromium, copper, manganese, molybdenum, nickel, vanadium, and zinc, ranged from 32.4 wt% for chromium to 94.7 wt% for molybdenum for a 3-hr leach in 8 N HCl at reflux temperature with 50% pulp density.[SO-14] Coal ash slag differs significantly from metal finishing industry wastes in that they contain aluminosilicate components in addition to mixed oxides and are heavily laden with a much wider distribution of elements other than the metals of interest for recovery.

Recent EPA contract studies have been conducted of the removal of metals from municipal sludges by hot acid treatments.[SO-15, SO-19, SO-31, SO-33, SO-40] Sulfuric acid treatments extracted from municipal sludge 80% or more of metals such as aluminum, zinc, manganese, nickel, and magnesium, 50 to 90% of chromium, and copper and lead to a lesser extent.[SO-19] Acid treatments of municipal sludge at 1.5 to 3 pH at 95°C produced solubilization on the order of between 88 and 100% for cadmium, 82 to 100% for zinc, 73 to 100% for nickel, and approximately 45% for chromium.[SO-19] The object of this latter study was sterilization and detoxification of the sludge rather than metal recovery.

Conventional municipal sewage sludges usually contain total nonferrous metal of less than 1%.[SO-15, SO-19] Landfills, notably those which have been recipients of metal-finishing industry sludges, can be expected to provide a much richer source of the metals, in many cases meriting consideration as ore beds for recovery.

Published studies are scarce for acid solubilization of industrial landfills or metal-finishing industry sludges with the object of metal recovery. A recent study at Montana College of Mineral Science and Technology at Butte[SO-22] shows the promise of efficient recovery of metals of interest, such as nickel, copper, zinc, and chromium, from acid solubilized waste metal sludges using combinations of solvent extraction and precipitation.

Acid treatment technology that has been developed for application to recovery from metal finishing industry wastes is the nitric acid treatment of cyanide electroplating wastes to precipitate the metals, as in the Bureau of Mines waste-plus-waste process.[SO-49]

Acid solubilization, notably sulfuric acid, has been applied for wastes ranging from petroleum ash,[SO-43] copper waste,[SO-41] various metal dusts,[SO-39] sludge ash,[SO-33] sewage sludge,[SO-31] and blast furnace slag and dusts.[SO-47] To a much lesser degree hydrochloric acid has been used with alloy scrap[SO-23, SO-48] and coal ash.[SO-35]

Alkali agents such as NaOH, NH_4OH, Na_2Co_3, and $NH_4OH/(NH_4)_2Co_3$ have been used as solubilizing agents for spent catalysts,[SO-44] magnetic tape,[SO-42] laterites,[SO-38] plating sludge,[SO-29, SO-30] and lead wastes.[SO-27] In some instances both acid and alkali solubilization stages have both been used.[SO-17, SO-22, SO-23, SO-28, SO-43]

Another alternative is to produce metal sulfates which have high water solubility by reaction with sulfates[SO-7, SO-8, SO-9, SO-25, SO-38] or indirectly from sulfides with subsequent oxidation to the sulfate.[SO-32]

Chlorination by reaction with chlorine compounds directly provides solubilization for some waste systems.[SO-23, SO-34, SO-44]

Reaction with carbon monoxide at temperatures on the order of 60 to 125°C for certain of the nonferrous metals such as nickel also provides a route for separation by formation of volatile nickel carbonyls.[SO-46]

And finally, certain of the biological agents provide solubilization in aqueous solution. The action of *Thiobacillus ferrooxidans* has been used commercially for heap leaching of copper from its ores. The biological systems have been discussed in Chapter 6 (see Table 6.1).

References

SO-1. Burkin, A. R. *Chemistry of Hydrometallurgical Processes* (Princeton, NJ: Van Nostrand Reinhold Co., 1966).

SO-2. Pehlke, R. D. *Unit Processes of Extractive Metallurgy* (New York: American Elsevier, 1973).

SO-3. Bautista, R. G. "Hydrometallurgy," in *Advances in Chemical Engineering*, Vol. 9, T. B. Drew et al., Eds. (New York: Academic Press, 1974).

SO-4. Habashi, F. "Hydrometallurgy," *Chem. Eng. News,* Feb. 8, (1982), pp. 46–58.

SO-5. Rule, A. R. and R. E. Siemens. "Recovery of Copper, Cobalt and Nickel from Waste Mill Tailings," *Proceedings of the 5th Mineral Waste Utilization Symposium*, U.S. Bureau of Mines and IIT Research Institute, Chicago, April 13–14 (1976).

SO-6. Paulson, D. L., et al. "Cobalt and Nickel Recovery from Missouri Lead Ores," *Proceedings of the 7th Mineral Waste Utilization Symposium*, U.S. Bureau of Mines and IIT Research Institute, Chicago, October 20–21 (1980).

SO-7. Joyce, F.E., Jr. "Extraction of Copper and Nickel from the Duluth Gabbro Complex by Selective High Temperature Sulfatization." U.S. Bureau of Mines RI-7475, PB 197026 (1971).

SO-8. Brooks, P. T. et al. "Improving Nickel Extraction from Oxide Nickel Ores." U.S. Bureau of Mines TPR 57, PB 213054 (1972).

SO-9. Joyce, F. E. Jr. et al. "Sulfatization Reduction of Manganiferrous Iron Ore." U.S. Bureau of Mines RI-7749, PB 221912 (1973).

SO-10. Stephenson, J. B., et al. "Recovery of Zinc Oxide from Galvanizing Wastes." *Proceedings of the 6th Min-*

eral Waste Utilization Symposium, U.S. Bureau Mines and IIT Research Institute, Chicago (1978).

SO-11. Tiwari, B. L. and D. D. Snyder. "Recovering Nickel from Zinc/Nickel Oxide Batteries," *Proceedings of the Symposium on the Nickel Electrode*, R. G. Gunther and S. Gross, Eds. Vol. 82-4. (Pennington, NJ: The Electrochemical Society, Inc., 1982).

SO-12. Twidwell, L. G. et al. "Industrial Waste Disposal. Excess H_2SO_4 Neutralization with Copper Smelter Slag," *Environ. Sci Technol.* 10:687 (1976).

SO-13. Golden, D. M. "Evaluation of Potential Processes for Recovery of Metals from Coal Ash." EPRI, CS-1992 Final Report (RP1404-2).

SO-14. Gilliam, T. M. and R. M. Canon. "Removal of Metals from Coal Ash" *AIAA Proceedings, 15th Intersociety Energy Conversion Engineering Conference*, Seattle, WA (August 18–22), Vol. 2 (1980), pp.970–977.

SO-15. Scott, D. W. "Removal and Recovery of Metals and Phosphates from Municipal Sewage Sludge." EPA-600/2-80-37 (1980).

SO-16. Smyres, G. A. and P. R. Haskett. "Recovery of Metal Values from Complex Sulfides." U.S. Patent No. 4410496 (1983).

SO-17. Gabler, R. C., Jr. and R. L. Stoll. "Removal of Leachable Metals and Recovery of Alumina from Utility Coal Ash." U.S. Bureau of Mines RI 8721 (1983).

SO-18. Hoffmann, J. E. "Recovering Platinum-Group Metals from Auto Catalysts," *J. Metals* 40(6):40 (1988).

SO-19. McNulty, K. J. et al. "Evaluation of Hot Acid Treatment for Municipal Sludge Conditioning." EPA-600/2-80-096 (1980).

SO-20. Darnay, A. and W. E. Franklin. "Salvage Markets for Materials in Solid Wastes." EPA SW-29C (1972).

SO-21. Alter, H. and W. R. Reeves. "Specifications for Materials Recovered from Municipal Refuse." EPA-670/2-75-034, PB 242540 (1975).

SO-22. Twidwell, L. G. "Metal Value Recovery from Metal Hydroxide Sludges." EPA 600/52–85/128, PB 86–157294 (1985).

SO-23. Davidson, C. F. "Extraction of Metals from Mixtures of Oxides of Silicates." U.S. Patent No. 4307066 (1981).

SO-24. Holman, J. L., Jr. and C. A. Neumeier. "Recovery of Metals from Grinding Sludges." U.S. Patent 4409020 (1983).

SO-25. Baglin, E. G. and J. M. Gomes. "Selective Recovery of Base Metals and Precious Metals from Ores." U.S. Patent No 4423011 (1983).

SO-26. Gabler, R. C., Jr. and J. R. Jones. "Metal Recovery from Secondary Copper Converter Dust by Ammoniacal Carbonate Leaching." U.S. Bureau of Mines RI 9199 (1988).

SO-27. Martin, G. O. "Extracting Copper from Lead-Copper Crusts." Romanian Patent No. 66599 (1978).

SO-28. Honda, A., E. Ishii, and Z. Inoue. "Treatment and Resources Recovery of Refuse Incinerator Dust," *Mizu Shori Gijutsu,* 19(12):1101–1116 (1978).

SO-29. Barna, V. and D. Dalalau. "Recovering Copper from Wastes and Residues by Ammoniacal Leaching." Romanian Patent No. 64193 (1978).

SO-30. Andersson, S. O. S. and M. J. Meixner. "Ammoniacal MAR Process—Recovery of Copper, Nickel and Zinc from Neutralization Wastes by Ammoniacal Leaching," *Aufbereit.-Tech.,* 20(5):264–268 (1979).

SO-31. Krofchak, D. "Treatment of Incinerated Sewage Sludge Ash." European Patent No. 4778 (1979).

SO-32. Spack, V., J. Beranek, and J. Tomasek. "Recovery of Copper by Leaching of Ore Waste." Czechoslovakian Patent No. 180132 (1979).

SO-33. Gabler, R. C., Jr. "Incinerated Municipal Sewage Sludge as a Potential Secondary Resource for Metals and Phosphorus." U.S. Bureau of Mines RI 8390 (1979).

SO-34. Frank, N. G. and I. Constantinescu. "Recovery of Iron, Copper, Zinc, Chromium and Nickel from Sludge of the Treatment of Wastewater from Electroplating." Romanian Patent No. 72211 (1980).

SO-35. Gabler, R. C., Jr. and R. L. Stoll. "Extraction of Leachable Metals and Recovery of Alumina from Utility Coal Ash." *Resour. Conserv.* 9:131–142 (1982).

SO-36. Sumitomo Metal Mining Co., Ltd. "Metal Recovery from Flue Dust Containing Copper." Japanese Patent No. 57201577 (1982).

SO-37. Dowa Mining Co., Ltd. "Metal Recovery from Industrial Wastes." Japanese Patent No. 58141346 (1983).

SO-38. Powers, L. A. and R. E. Siemens. "Examination of Effluents Generated from Processing Domestic Laterites." U.S. Bureau of Mines RI 8797 (1983).

SO-39. Wojtowicz, J., Z. Smieszek, J. Szymanski, T. Padewski, W. Rybak, L. Gotfryd, Z. Szolomicki, A. Madejski, and Z. Myczkowski. "Recovery of Metals from Zinc- and Copper-Containing Polymetallic Dusts." Polish Patent No. 121761 (1983).

SO-40. Haas, O. W. "Sludge Demetalization by the Union Carbide Corporation Electrochemical Process." EPA/600/2-84/196; PB85–137347/GAR (1984).

SO-41. Zakrzewski, J., J. Wojtowicz, E. Lach, A. Chmielarz, and G. Benke. "Method for Producing Ammonium Rhenate from Washing Sulfuric Acid Used in Copper Production," *Przem. Chem.* 64(3):143–146 (1985).

SO-42. Mitsumaru Kagaku K. K.; Sony Corp. "Treatment of Waste Magnetic Tape." Japanese Patent No. 60096733 (1985).

SO-43. Akaboshi, T., M. Tsuruta, A. Sakuma, M. Tijima, K. Sato, and N. Saneko. "Treating of Ash from Combustion of Petroleum-Derived Fuels." Japanese Patent No. 61171582 (1986).

SO-44. Matsuda, K. "Recovery of Metal Values from Spent Catalyst Derived from the Desulfurization of Crude Oil," *Suiyokaishi* 20(4):254–261 (1985).

SO-45. Goto, S., T. Hashimoto, Y. Marukawa, G. Nada-zawa, and T. Ito. "Recovery of Valuable Metals in Manganese Nodules." Japanese Patent No. 61207528 (1986).

SO-46. Farkas, I., R. Csikos, J. Bathory, L. Panyor, and F. Nagy. "Process for Recovering Nickel from Catalysts and Wastes Containing Nickel in Nickel Tetracar-bonyl Form." Hungarian Patent No. 39632 (1986).

SO-47. Christian, J. and I. Joseph. "Recovery of Metals from Wastes by Acid Leaching." Belgian Patent No. 1000323 (1988).

SO-48. Laverty, P. D., G. B. Atkinson, and D. P. Desmond. "Separation and Recovery of Metals from Zinc-Treated Superalloy Scrap." U.S. Bureau of Mines RI 9235 (1989).

SO-49. Cochran, A. A., et. al. "Development and Application of the Waste-Plus-Waste Process for Recovering Metals from Electroplating and Other Wastes." U.S. Bureau of Mines RI 7877 (1974).

8

Metal Recovery from Multimetal Wastes

Multistage separation schemes need to be designed as the most appropriate for a given multimetal waste system. Separation schemes can be assembled from selection of the best unit processes such as described in Chapters 5 to 7, with consideration given to the chemical composition of the waste and the economics of available technology.

Examples are presented here of separation schemes devised for two kinds of frequently complex multimetal wastes: metal finishing industry sludges and spent catalysts.

8.1 METAL FINISHING INDUSTRY WASTES

Metal finishing industry mixed-metal hydroxide sludges arise principally from cleanup of wastewater to meet EPA effluent emission standards or from disposal of spent process solutions. The metals commonly encountered consist of iron, aluminum, chromium, cadmium, cobalt, copper, manganese, molybdenum, nickel, lead, tin, titanium, and zinc, and in addition, other metals such as niobium, tantalum, zirconium, etc. Waste sludges commonly will contain up to six of these metals, but only one or two may merit recovery rather than safe disposal.

One approach to separation and recovery for several of the metals present in significant amount in complex metal hydroxide wastes has been evaluated at laboratory scale and in addition at pilot plant scale with a processing rate of 75 to 100 pounds of sludge per day by Twidwell.[MF-1] Processing stages consisted of (1) sulfuric acid solubilization, (2) iron separation by jarosite precipitation, (3) copper separation by solvent extraction, (4) cadmium and zinc separation by solvent extraction, (5) chromium (Cr^{6+}) separation by precipitation as lead chromate, and a final stage separation of nickel by one of several alternatives, such as crystallization.

An alternative separation scheme evaluated by Twidwell and Dahnke[MF-2] consisted of removal of the trivalent contaminant metal cations Al^{3+}, Cr^{3+}, and Fe^{3+} by phosphate precipitation in a moderately acid pH range to facilitate the efficiency of separation of the more valuable divalent metal cations Cu^{2+}, Cd^{2+}, Ni^{2+}, and Zn^{2+}.

An adaptation of this Twidwell scheme[MF-3] using waste compositions and production volumes for a single state, Missouri, has been used to make an economic assessment for application to a state centralized treatment plant for metal wastes, leading to promising results. This conceptual plant designed for processing 16,890 metric tons of metal waste per year, built at a fixed capital cost (1982) of $5,580,000, recovered total metals valued at $870,000 per year. Flowsheets for this conceptual plant are shown in Figures 8.1 and 8.2.

Another multistage separation scheme using adaptations of known technology to treat mixed-metal finishing industry wastes is that of Recontek at a plant which went on stream in 1990 at Newman, Illinois. Typical feed stock consists of multimetal wastes, principally EPA designations F006, K061, K062, D002, D008, D009, and D011, as mixed acids and alkalies, dry and wet sludges. Separation stages, selected to be appropriate for the waste composition, consist of removal of chromium and/or iron by phosphate precipitation, electrowinning separation of copper and zinc, and nickel separation by crystallization as nickel sulfate.[MF-4]

The Newman, Illinois, plant has a capacity of 6,000 tons per month. The principal incentive for a generator with

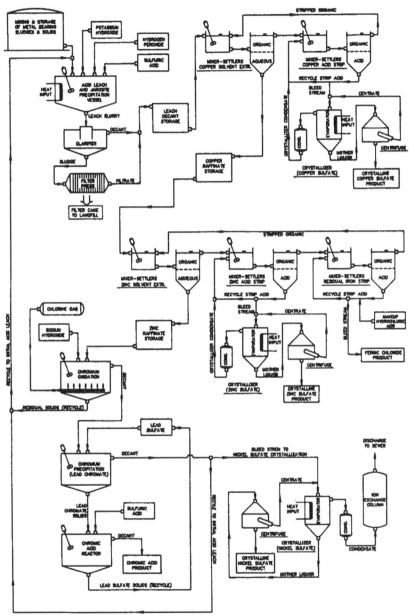

Figure 8.1 Simplified process flow for treatment of metal finishing wastes (adaptation of Twidwell processes). (From Reference MF-3, page 690.)

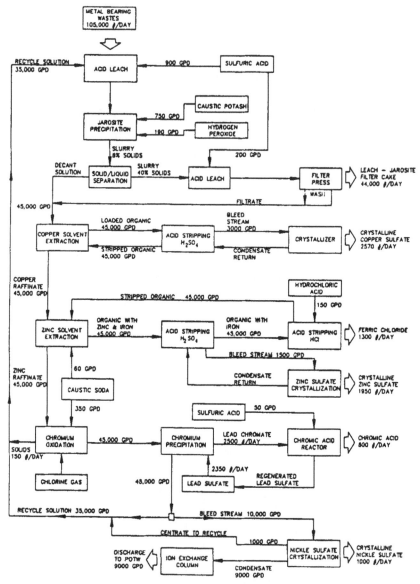

Figure 8.2 Mass balance flowsheet for treatment of metal finishing wastes (adaptation of Twidwell process). (From Reference MF-3, page 698.)

metal wastes falling within the EPA hazardous waste categories is to minimize contingent liability by a process which
detoxifies by metal recycling and minimizing landfill residues. Credits obtained for the recovered metals are shared
with the generator, making the overall cost of waste disposal significantly lower than conventional landfill disposal. Plans to build similar plants in other states are in
progress.

Another separation scheme for recovery of copper and
nickel from metal finishing wastes, using primarily precipitation and solvent extraction processes, has been demonstrated to be technically feasible by bench scale experimental
evaluation with industrial wastes.[MF-5, MF-6] A flow scheme for
systems containing chromium, iron, copper, nickel, and
zinc, with chromium and iron at contaminant concentrations
not in excess of 1,000–5,000 ppm, is shown in Figure 8.3.
Following solubilization of any solid sludges in sulfuric acid,
adjustment with alkali to pH ~ 2 to 3.5 precipitates most of
the chromium and iron; solvent extraction removes the
residual iron and the copper; an oxidation stage converts the
Cr^{3+} to Cr^{6+} and destroys the organics (metal complexing
agents and residual solvent extraction agents and solvent),
and the nickel is precipitated as nickel oxalate. Ion exchange
can be used as necessary to complete the nickel recovery.
Residual metals in the effluent can be reduced by sulfidation
and/or adsorption to meet EPA emission standards. Results
for five high-nickel content wastes, electrochemical milling
filter cake, nickel hydroxide sludge, waste pickling acid, and
a nickel catalyst and a spent electroless nickel waste are
shown in Table 8.1 and show that nickel can be separated
with an efficiency of 93 to 99%.

A modification of this separation scheme can be used for
metal wastes with chromium, iron, copper, and nickel,
where chromium and iron are present at high levels (> 1,000
to 5,000 ppm) and where it is desired to recover both copper
and nickel. In this modification, the chromium and iron in
the trivalent state can be removed by phosphate precipitation after acid solubilization prior to the copper and nickel

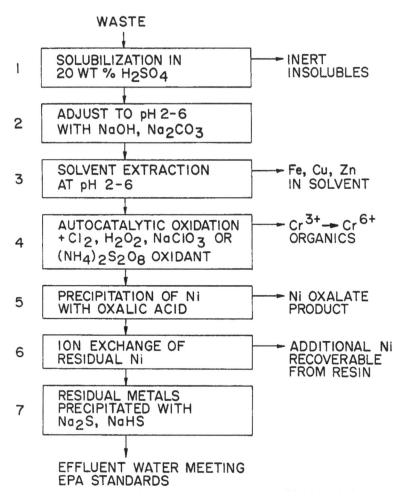

Figure 8.3 Schematic for multistage separation of high nickel content industrial wastes.

separation stages. Subsequent stages consist of copper separation by solvent extraction and nickel recovery by oxalate precipitation and ion exchange as indicated in the unmodified scheme (Figure 8.3).

Table 8.1 Recovery of Nickel from Staged Separation of Industrial Wastes

			Summary of separation results		
Waste	Waste nickel content (wt%)	Separation stages	Nickel wt% recovered by oxalate precipitation	Nickel wt% recovered by ion exchange	Residual nickel (ppm)
ECM	4.4	1,2,3,5,6	99.0	0.88	3.5
Nickel hydroxide	5.7	1,2,3,5,6	96.9	3.0	6.0
Mineral acid	3.0	2,3,5,6	96.9	3.1	10.0
Nickel catalyst	25.0	1,2,5,6	99.9	0.07	18.0
Electroless nickel	0.29	4,6	92.5	2.2	217.0

References

MF-1. Twidwell, L. G. "Recovery from Hydroxide Sludges: Final Report." EPA 600/2–85/128; NTIS PB 86157294 (1984).

MF-2. Twidwell, L. G. and D. R. Dahnke. "Metal Value Recovery from Metal Hydroxide Sludges: Removal of Iron and Recovery of Chromium." EPA 600/2–88/019. P988–176078 (1988).

MF-3. Ball, R. O., P. L. Buckingham, and S. Mahfood. "Economic Feasibility of a State-Wide Hydrometallurgical Recovery Facility," in *Metals Speciation and Recovery*, J. W. Patterson and R. Passino, Eds. (Chelsea, MI: Lewis Publishers, Inc., 1987).

MF-4. Rajcevic, H. P. "A Hydrometallurgical Process for Treatment of Industrial Wastes," *Plating and Surface Finishing*, July 22 (1990).

MF-5. Brooks, C. S. "Nickel Metal Recovery from Metal Finishing Industry Wastes," *Proceedings of the 42nd Industrial Waste Conference*, Volume Date 1987. (Chelsea, MI: Lewis Publishers, Inc., 1988), pp. 847–852.

MF-6. Brooks, C. S. "Metal Recovery from Electroless Plating Wastes," *Met. Finish.* 87(5):33–36 (1989).

M-7. Borner, A. J. and B. Perry. *Hazmat World* (November 1990), p. 40.

Table 8.2 Metal Recovery from Spent Catalysts

Catalyst	Metals	Pyrometallurgy	Leaching
Catalysts electrical and electronic	Ag, Cu	Heat to $<1400°/O_2/$ Fe Slag	–
Co/Mo petroleum catalyst	Co, Mo, Ni, V	Roast with alkali agent	Water/alkali leach
Co/Mo petroleum catalyst	Al, Co, Mo, Ni, V	Heat with sulfates to 220°C	–
Ni petroleum catalyst	Ni	Roast to 600°C + melt with Fe ore	H_2SO_4 leach to 70°C
Desulfur catalyst	Co, Mo, Ni, V	Roast to 1173°K with Na_2CO_3	Hot water leach
Ni catalyst	Ni	H/NH_3/syn gas. to 450°C + sulfide	CO reduction 125°C
Ni/Fe contaminated FCC catalyst	Ni	–	–
Petroleum catalyst	Co, Mo, Ni, V, Al	Late stage steam to 750°F	HCl leach
Petroleum catalyst	Al, Co, Mo, Ni, V	Roast to 700°	HCl leach
Oil hydrogen catalyst	Ni	Final stage formate reduction	HCl/HNO_3/H_2SO_4
Raney Ni, Ni/Al	Ni	–	Leach with electrolyte with N_2, Cu, Te, Mg, H_2SO_4
Ni catalyst	Ni	Heat to 150° with NH_3/O_2	–

8.2 METAL RECOVERY FROM SPENT CATALYSTS

Spent catalysts provide additional significant sources for recovery of nonferrous metals from industrial wastes. The U.S. catalyst market of $1.8 billion per year represents 60% of the world market with applications of 37% in petroleum refining, 34% in chemical processing, and 29% in emission control and a projected annual growth rate of 4.5% or more SC.[SC-26, SC-27]

Table 8.2 continued

Subsequent hydrometallurgical processing	Metal recovery efficiency %	Ref.
Terminated with Cu/Ag alloy product	76.5 Cu/91.3 Ag	SC-2
Magnetic separation Ni/Co oxides	–	SC-3
Co/Mo/Al in solution + heat treat to form Al_2O_3	97 Co, 92 Mo	SC-4
26.3 gr Ni/L in solution	96 Nickel	SC-5
H_2 reduction to 1443°K + $FeCl_3$ leach	>90 Co/Ni	SC-6
Recovery Ni carbonyl	92–95 Ni	SC-7
High-gradient magnetic separation	Ni/Fe/cat	SC-8
Solvent extraction, phosphate precipitation		SC-9
Solvent extraction, precipitation	Co, Mo, Ni, V	SC-10
NaOCl, hydroxide precipitation	Ni catalyst reactivated	SC-11
Catalyst added to precipitation Fe/Cu/Al	$Ni(SO_4)\cdot 7\ H_2O$ product	SC-12
Ni amine complex formed	Ni as metal, oxide, salt	SC-13

The applications of these catalysts fall into several major categories: hydrotreating of petroleum and coal fuels with metals such as Co, Mo, W, Ni; hydrogenation and oxidation of a wide range of CPI organics over copper, nickel, chromium, and zinc metals in various combinations; steam refining over nickel; automotive exhaust cleanup with platinum and rhodium; petroleum reforming over platinum; shift conversion of carbon monoxide over combinations of copper, chromium, iron, and zinc; fuel combustion over platinum in

Table 8.2 continued

Catalyst	Metals	Pyrometallurgy	Leaching
Automotive catalyst	Pt, Pd	Possible pre-reduction H_2/N_2 to 300°	HCl/HNO$_3$
Cr/Fe shift catalyst	Cr, Fe	Roast with NaOH or Na$_2$CO$_3$	H$_2$O leach
Hydroprocessing catalysts	Cu, Mo, Ni, V	Chlorination to 450°	NaOH/H$_2$SO$_4$ leach
Automotive catalysts	Pt, Pd, Rh	Chlorination to 700° or plasma fusion to 2000°	H$_2$SO$_4$ leach
Catalyst leachate	V	Heat to 1000° with air/Na, K salts	–
Hydroprocessing catalysts	Cr, Cu, Mo, Ni, W	Chlorination to 450°	H$_2$SO$_4$/NaOH leach
Polyphenylene oxide catalyst	Cu	–	–
Hydrogenation catalyst	Pd	Heat to 90° in air	NH$_4$Cl/NH$_4$ NO$_3$
Reforming catalyst	Ni	–	Alkali
Automotive catalyst	POt	Plasma vaporization	–
Petroleum HDS/coal liquefaction catalyst	Co, Mo, Al	Roast/air/Na$_2$CO$_3$/750°	NaOH leach
Petroleum catalyst for residual fuel	V	Roast	Hot alkali leach

fuel cells, vegetable oil hydrogenation over nickel, and a wide number of various metals used in the chemical process industries for diverse synthesis reactions.

Catalytic metals used in large enough volume to merit serious consideration for recovery rather than disposal are the hydrotreating catalysts, which consist of combinations of cobalt and nickel with molybdenum or tungsten, nickel catalysts used in steam refining and vegetable oil hydrogenation, and the platinum catalysts along with rhodium used in petroleum refining, fuel cells, and automotive catalysts.

Table 8.2 continued

Subsequent hydrometallurgical processing	Metal recovery efficiency %	Ref.
Prereduction not recommended	74–98 Pt, 74–96 Pd	SC-14
Precipitation, calcine, membrane separation	90–99 of Cr_2O_3	SC-15
Solvent extraction, precipitation	36–99 metal recovery	SC-16
Cl_2/SO_2 reaction, cementation, precipitation	80–97 Pt, 80–96 Pd, 60–95 Rh	SC-25
Hydroxide precipitation	90 V as V_2O_5	SC-17
Solvent extraction, electrowinning	–	SC-18
Liquid-liquid extraction, distillation	–	SC-19
N_2H_4 reduction	90 Pd	SC-20
Crystal as Ni salt	–	SC-21
Condensation metal vapors	–	SC-22
Precipitation Mo, Co/Al_2O_3	95 Mo	SC-23
Precipitation/extraction V for separation from Mo	Separation V and Mo	SC-24

Petroleum processing catalysts also provide a source for metals in addition to those initially present as catalysts, namely, the nickel and vanadium accumulated from reaction with the organometallics in the petroleum feed stock.

A number of published results for a variety of metal catalysts are summarized in Table 8.2. The processing steps fall into three principal stages: (1) pyrometallurgical reduction to enhance solubilization, (2) solubilization by acid or alkaline leaching, and (3) various hydrometallurgical separation pro-

Table 8.3 Companies Treating and/or Reclaiming Metal Waste

Company	Location	Metal treatment or reclamation
Climax Molybdenum (AMAX)	Ann Arbor, MI	Hydrotreating catalysts
CRI-MET	Houston, TX	Petroleum catalysts
Handy & Harman	New York, NY	Precious metals
PYP Industries Inc. (Gerald Metals)	Stamford, CT	Precious metals
CP Chemicals	Sewaren, NJ	Metal salts
ETICAM	Warwick, RI	Metal finishing waste
UNC Reclamation	Mulberry, FL	Metal finishing waste
Recontek	San Diego, CA	Metal finishing waste
Envirite	Plymouth Meeting, PA	Metal finishing waste
The MacDermid Group	Plymouth, CT	Metal finishing waste
Met-Pro Systems Division	Harleysville, PA	Metal finishing waste
UNOCAL	Schaumburg, IL	Metal finishing waste
SERFILCO, Ltd.	Glen View, IL	Metal finishing waste
Encycle (Asarco)	Corpus Christi, TX	Metal finishing waste
EWR	Waterbury, CT	Metal finishing waste

cesses such as electrowinning, ion exchange, precipitation, solvent, extraction, etc.

A number of companies conduct treatment and in some cases metal reclamation for metal finishing industry wastes, and several companies specialize in spent catalysts (Table 8.3). Companies such as ETICAM, Encycle, Recontek, and UNC Reclamation process complex mixed-metal wastes, principally from the metal finishing industries and feature metal reclamation. Two companies, Climax Molybdenum and CRI-MET, specialize in metal reclamation from petroleum hydrotreating catalysts. Companies such as Handy & Harman and PYP Industries specialize in reclamation of precious metals from metal wastes.

References

SC-1.   Hennion, F. J. and J. Farkas. "Assessment of Critical Metals in Waste Catalysts." U.S. Bureau of Mines OFR 197–82 PB83–144832 (1982).

SC-2.   Van Hecke, M. C. F. and L. M. Fontainas. "Extraction of Nonferrous Metals from Residues Containing Iron." European Patent 53406 (1982).

SC-3.   Mitsubishi Steel Mfg. Co., Ltd. "Recovery of Valuable Substances from Cobalt-Molybdenum Waste Catalysts." *Jpn. Kokai Tokkyo Koho*, Japanese Patent JP 55089437 (1980).

SC-4.   Rastas, J. K., K. J. Karpale, and H. Tiitinen. "Recovery of Metal Values from Spent Catalysts Used in Extracting Sulfur from Crude Petroleum." Belgian Patent BE 894678 (1983).

SC-5.   Zambrano, A. R. "Recovery of Nickel from Wastes." German Patent DE 3000040 (1980).

SC-6.   Matsuda, K. "Recovery of Metal Values from Spent Catalyst Derived from the Desulfurization of Crude Oil," *Suiyokaishi* 20(4):254–261 (1985).

SC-7.   Farkas, I., R. Csikos, J. Bathory, L. Panyor, and F. Nagy. "Process for Recovering Nickel from Catalysts and Wastes Containing Nickel in Nickel Tetracarbonyl Form." Hungarian Patent HU 39632 (1986).

SC-8.   Takase, S., M. Ushio, Y. Oishi, T. Monota, and T. Shiori. "Magnetic Separation of Fcc Equilibrium Catalyst by HGMS," *Am. Chem. Soc. Div. Pet. Chem. Prepr. Symp. on Recovery of Spent Catalysts* 27(3):697 (1982).

SC-9.   Sefton, V. B., R. Fox, and W. P. Lorenz. "Recovery and Separation of Metal Values from Spent Petroleum Refining Catalyst." U.S. Patent US 4861565 (1989).

SC-10. Ward, V. C. "Meeting Environmental Standards when Recovering Metals from Spent Catalyst," *JOM* 41(1):54–55 (1989).

SC-11. Shukla, A., P. N. Maheshwari, and A. K. Vasishtha. "Reclamation of Nickel from Spent Nickel Catalyst." *J. Am. Oil Chem. Soc.* 65(11):1816–1823 (1988).

SC-12. Charewicz, W. A., T. Chmielewski, B. Kolodziej, and J. Wodka. "Hydrometallurgical Recovery of Nickel from Spent Catalysts," *Rudy Met. Niezelaz.* 32(2):61–65 (1987).

SC-13. Tamai, Y., T. Okabe, and H. Ishii. "Recovery of Nickel from Residues." German Patent DE 2808263 (1978).

SC-14. Tyson, D. R. "Leaching of Platinum and Palladium from Spent Automotive Catalysts," M.S. Thesis, Iowa State University, Ames. (1984).

SC-15. Stubbs, A. M. and B. W. Song. "Chromium Recovery from High Temperature Shift Cr-Fe Catalysts." U.S. Bureau of Mines RI 9204 (1988).

SC-16. Jong, B. W., S. C. Rhoads, A. M. Stubbs, and T. R. Stoelting. "Recovery of Principal Metal Values from Waste Hydroprocessing Catalysts," U.S. Bureau of Mines RI 9252 (1989).

SC-17. Rodriguez, D., R. Schemel, and R. Salazar. "Precipitation or Recovery of Vanadium from Liquids." German Patent DE 3509372 (1985).

SC-18. Jong, B. W. and R. E. Siemens. "Proposed Methods for Recovering Critical Metals from Spent Catalysts," in *Recycle Second. Recovery Met., Proc. Int. Symp.*, P. R. Taylor, H. Y. Sohn, and N. Jarrett, Eds. (Warrendale, PA: Metallurgical Society, 1985), pp. 477–488.

SC-19. Ciernik, J., E. Sousta, M. Stastny, D. Ambros, L. Preisler, J. Sramek, and P. Rysanek. "Copper Recovery with Regeneration of Liquors after Oxidation Polycondensation of 2,6-Xylenol." Czechoslovakian Patent CS 250949 (1988).

SC-20. Heves, A. "Palladium Recovery by Leaching of Spent Hydrogenation Catalysts." Romanian Patent RO 96884 (1987).

SC-21. Deb, K. B., R. M. Sanyal, S. K. Ghosh, J. L. Bhowmick, and K. R. Chakravorty. "Recovery of Nickel in the Form of Nickel Salts from the Effluent in the Manufacture of Nickel Based Reforming Catalysts." Indian Patent IN 139459 (1976).

SC-22. Tajmouati, A., G. Peraudeau, J. M. Badie, G. Vallbona, B. Bonduelle and J. P. Guerlet. "Plasma Vaporization: Application to the Recovery of Precious Metals from Spent Automotive Exhaust Catalysts," *Colloq. Phys.*, C5, Eur. Congr. Therm. Plasma Processes Mater. Behav. High Temp (1990).

SC-23. Sebenik, R. F. and R. A. Ference. "Recovery of Metal Values from Spent CoMo (Al_2O_3) Petroleum Hydrodesulfurization and Cool Liquefaction Catalysts = Laboratory-Scale Process and Preliminary Economics," *Am. Chem. Soc. Div. Pet. Chem. Prepr. Symp. on Recovery of Spent Catalysts* 27(3):674 (1982).

SC-24. Nevitt, T. D. and R. S. Bertolini. "Solvent Extraction of Vanadate Anion from the Alkali Leachate of Used Resid. Catalysts," *Am. Chem. Soc. Div. Pet. Chem. Prepr. Symp. on Recovery of Spent Catalysts* 27(3):683 (1982).

SC-25. Hoffman, J. E. "Recovering Platinum-Group Metals from Auto Catalysts," *J. Met.* 40(6):40(1988).

SC-26. "Process Catalysts Continue on Steady Course," *Chem. Eng. News*, Feb. 17 (1986), p. 27.

SC-27. "Catalysts Reacting to Changing Markets," *Chem. Week* 148(24):34 (1991).

9

Metal Recovery Economics[1]

9.1 INTRODUCTION

The total volume of hazardous waste generated in the United States during 1986 was 747 million tons.[2] Electroplating wastewater treatment sludge, or F006 waste,[3] is a waste stream commonly found recycled today. In 1986, F006 waste accounted for 11.2 million tons, or 1.5% of the total volume.[4] Our investigations tell us that a metal recovery plant can process in the range of 50,000 to 100,000 tons of waste per year. Thus, it would take 224 to 448 plants to treat the F006 waste stream alone. At present, only a handful of plants treat this waste stream. Enormous potential, therefore, exists for the recovery of metals from this and other types of metal-bearing wastes.

Metal recovery, as highlighted in this book, is indeed technologically feasible for a number of waste streams. However, not all recovery processes make sense in light of the economics of the chosen process. The purpose of this chapter is to outline the important issues that one should consider when evaluating the economic feasibility of a metal recovery project. Our first task is to define the markets for metal recovery. Next, we identify key factors which contribute to the economics of the different markets. As expected, many of these factors change in relative importance depending on the market situation of a metal recoverer. To illustrate this phenomenon, we examine a few prevalent scenarios in which a metal recovery operation may make economic sense. Additionally, actions that the government could take to improve the economic feasibility of metal recovery are summarized and a list of potential data sources is included.

Our discussion throughout the chapter is qualitative in nature. This is not to say that rigorous financial analysis is not needed to accurately assess the economic feasibility of a metal recovery venture. Rather, our chapter raises the most important issues that should be considered for such an analysis and provides a basic context for illustration of these effects. Perhaps the most important service we can provide is a list of data sources. Since the technology of metal recovery is changing rapidly, our analysis may become outdated quickly. The data sources we provide in Appendix B will help a potential metal recoverer find the necessary, up-to-date information required to perform accurate analyses.

9.2 MARKET DEFINITION

The metal recovery market evolves out of the fact that many industrial processes produce metal-bearing wastes. There are only two options for managing such wastes: disposal, either with or without treatment, or recycling. The choice of disposal vs recycling is primarily an economic one. A generator of metal-bearing wastes will choose whichever method costs the least. Although simple in principle, this choice can be fairly complicated, especially in the presence of heavy government regulation. In fact, government regulation plays the single most important role in defining the market for metal recovery. Estimating the cost implications of regulation is the most difficult aspect of evaluating disposal and recycling options.

As recently as the early 1980s, disposal of metal-bearing wastes was commonplace and inexpensive in comparison to metal recovery. The Resource Conservation and Recovery Act of 1976 (RCRA) and the Comprehensive Environmental Response, Compensation and Liability Act of 1980

(CERCLA), or Superfund, as amended by the Superfund Amendments and Reauthorization Act of 1986 (SARA), have each contributed to a rise in disposal costs to the generator, increasingly making metal recovery an economically attractive alternative. While the second chapter in this book outlines the regulations governing metal recovery, this section specifically addresses the effects that recent regulatory trends have had on metal recovery economics.

RCRA is the main driving force behind the recent rise in metal waste disposal fees. Authorized in 1976 and amended in 1984 through the Hazardous and Solid Waste Amendments (HSWA), RCRA maintains the following goals:

- to move away from land disposal as the primary means of hazardous waste management by requiring treatment of wastes before final disposal
- to reduce the environmental and health risks at land disposal facilities
- to close down facilities that cannot safely manage waste
- to close loopholes in the types of waste management facilities not covered under RCRA[5]

Many metal wastes were not initially affected by RCRA as a result of the Bevill Amendments (Amendments). These Amendments, passed in 1980, required EPA to determine whether a waste is actually hazardous before it comes under the jurisdiction of RCRA.[6] Under these Amendments, mine wastes were initially exempt from regulation.[7] However, the EPA approved a final rule establishing 17 metal mining wastes as hazardous under RCRA (see Table 9.1 for a list of these wastes).[8] Treatment and/or landfilling of these wastes is now required, taxing the already shrinking national Treatment, Storage and Disposal Facility (TSDF) capacity. The EPA estimates that the inclusion of these wastes under RCRA's jurisdiction will cost the mining industry approximately $53 million per year in RCRA compliance alone (this does not include the resulting increase in CERCLA costs).[9]

Recent amendments promulgated under RCRA have made disposal even more costly.[10] The so-called "land ban" restricts

Table 9.1 17 Waste Streams Recently Regulated Under RCRA

1. Barren filtrate and processing raffinate from beryllium smelting
2. Process water from cerium production
3. Cooling tower blowdown from coal gasification
4. Acid plant scrubber blowdown from copper smelting
5. Bleed electrolyte from copper smelting
6. Process waste water from copper smelting
7. Furnace scrubber blowdown and process waste water from the production of elemental phosphorus
8. Ammonium nitrate process solution from lanthanide production
9. Acid plant blowdown from primary lead smelting
10. Scrubber waste water from light-weight aggregate production
11. Selenium plant effluent from processing acid plant blowdown in the production of molybdenum
12. Air pollution control scrubber blowdown from tin production
13. Chloride processing waste acids and leach liquor from titanium production
14. Acid plant blowdown from primary zinc smelting
15. Process waste water from primary zinc smelting
16. Soda ash from trona ore processing
17. Bertrandite thickener slurry from beryllium production

Note: All waste streams are covered under RCRA Subtitle C except numbers 16 and 17, which are covered under Subtitle D.

land disposal and requires treatment by an accepted technology for many waste streams. A significant number of metal-bearing waste streams are regulated under this rule. The effect of restricting land disposal and forcing additional treatment of metal-bearing wastes has driven disposal costs for generators upwards. Add to this the decrease in the number of land disposal sites (as of 1990, 87% of RCRA-permitted landfill facilities that were in existence at HSWA's inception are either closed or in the process of closing)[11] and the result is an escalation in the cost of disposing of hazardous waste. For example, over the period 1980 to 1987, average tipping fees[12] at hazardous landfills increased from $50[13] to over $130 per ton.[14] As this trend toward increasing disposal costs continues, metal recovery becomes a more economically attractive means of handling hazardous waste.

Table 9.2 Schedule of Superfund Waste Tax

Year	Land disposal tax (dollars per ton)
1986	$37
1987	39
1988	42
1989	44
1990	47

Source: Superfund Amendments and Reauthorization Act of 1986, U.S. House of Representatives, Report 99–962, October 3, 1986, p. 331.

While RCRA contributed to the increase in the disposal fees that a generator faces, CERCLA increased the potential liability that a generator assumes whenever waste is disposed of in such a way as to pose a threat to the environment. Superfund was originally authorized by Congress in 1980 to provide the financial means for the EPA to clean up extremely hazardous waste sites. CERCLA assigned the liability for cleanup not only to the owner/operator of the site, but also to the generators of the wastes that were treated and disposed of at the site and the transporters who brought the waste to the site. The Superfund program is expected to continue well into the next century. At present, there are 1,187 sites on the National Priorities List (NPL),[15] with over 30,000 potential sites identified.[16] With a current expected cost for cleanup of $31.6 million per site,[17] the expected total liability is $37.5 billion for the present NPL sites alone.

By holding generators liable for the cleanup of hazardous waste sites, the federal and state governments have basically added an additional cost for every ton of material that is sent to a landfill—a contingent liability. In fact, under SARA, the EPA has begun levying a tax for every ton of hazardous waste disposed of in a landfill to recover costs not covered by generators (see Table 9.2 for the Superfund tax schedule for the period 1986 to 1990). Virtually every metal recoverer we interviewed mentioned generator contingent liability as one of the main reasons that generators are interested in recycling as a method of handling their wastes. Sources cited

Table 9.3 Estimation of Contingent Liability Associated with Landfill Disposal

This calculation is performed for two periods, 1984 and 1989, in order to reflect any escalation that has taken place. These numbers are best estimates and are used solely to determine the scale of the number.

	1984	1989
1. Average cleanup cost of a Superfund site: (millions of dollars)	$8.84[a]	$31.57[b]
2. Average volume at a Superfund site:[c] (cubic yards)	822,608	822,608
3. Average contingent liability: (dollars per cubic yard − (1)/(2))	$11	$38

Note: 1 cubic yard ~ 1 ton.

[a]"Extent of the Hazardous Release Problem and Future Funding Needs: CERCLA Section 301(a)(1)(C) Study," U.S. Environmental Protection Agency, Office of Solid Waste and Emergency Response (December 1984).

[b]40 CFR Part 300, "National Priorities List for Uncontrolled Hazardous Waste Sites" (August 30, 1990), p. 23.

[c]Derived by taking average volume of waste at all landfill sites (63) in the "ROD Annual Report: FY 1989", pp. 151–297. Note that volumes ranged from 90 cubic yards to 27 million cubic yards per site, illustrating the uncertainty of relying on a single point estimate for decision making.

Additional considerations:

1. Probability that the landfill being used will become a Superfund site. Not all landfills will require remediation.
2. The expected amount of time until cleanup is required. The contingent liability is discounted to arrive at a present value estimate.
3. Risk averseness. The aforementioned discounting of the contingent liability will vary depending on a generator's risk profile (i.e., a risk-averse generator may pay a premium above the total contingent liability if it means avoiding the risk of the liability altogether).

perceived contingent liability premiums ranging from $40 to $300 per ton for solid wastes and sludges[18] and as much as two to three times the treatment cost for metal-bearing liquid wastes.[19] They noted that some generators will pay this additional premium for recycling above the landfill disposal fees in an effort to avoid the associated contingent liability from landfilling.

Understandably, there still remains a great deal of uncertainty as to the magnitude of a typical company's contingent liability for waste landfilling. Calculations of average contingent liabilities per ton of waste disposed for 1984 and 1989 appear in Table 9.3. Due mainly to the growth of Superfund cleanup costs over the period, the estimated contingent liability grew from $11 to $38 per ton of waste landfilled. Performing similar calculations for a given generator or for a given waste stream would require very specific information and a more detailed analysis which is beyond the scope of this book. But the results from both this rough calculation and the perceived liability premium raised by industry sources show that a contingent liability in terms of dollars is significant.

In addition to significantly increasing the disposal fees that a generator must pay to a landfill, RCRA has also complicated the operating environment of a metal recoverer. In a paper analyzing the effect of RCRA on metal recovery efforts, the Bureau of Mines (BOM) observes that, "The EPA has interpreted the definition of solid waste very broadly to include all but a few recycling situations."[20] The BOM continues, indicating that the EPA has exempted at least one recycler from regulation. This determination, however, is made on a case-by-case basis.[21] The BOM writes with reference to RCRA limiting the recycling and reuse of spent aluminum potliner:

Much of the potential cost increases from RCRA regulating recycling result from two rules which were being imposed on the reuse of spent potliner, the "derived-from" and the "mixture" rules. These two measures require that wastes derived from hazardous wastes or mixed with hazardous wastes are to be treated as hazardous wastes. This means, in the case of recycled spent potliner, that all of the residual waste from the reuse processes, such as the coal ash from a cement kiln, would be considered hazardous waste, even though only 2 percent of the total coal input consisted of spent potliner and the hazardous constituent, cyanide, is destroyed in the process.[22]

The BOM concludes that:

The present framework for controlling wastes incorporates the regulation of recyclable materials, providing in some instances, questionable, if any, additional benefits to society. If this trend continues, regulations could lead to the end of major recycling efforts, such as scrap iron, lead-acid batteries, or aluminum itself, which are essential to our society and extremely beneficial to the environment.[23]

As of this writing, Congress is scheduled to clarify the intent of RCRA for recyclers in its 1991 session.[24]

While operating within RCRA is not impossible, it can increase the operating costs of a recycling operation through additional reporting and other compliance expenses. A facility that is currently not under the jurisdiction of RCRA would in all likelihood want to avoid being regulated because of the additional regulatory costs and scrutiny that would ensue. In fact, marginally profitable ventures could be forced out of business by the additional costs of compliance.

The effect of Superfund on a metal recovery operation itself is more difficult to assess. Beyond the obvious case of sham recyclers,[25] the question here is whether a recycled product (e.g., copper or zinc sulfate) purchased by a waste-producing facility that later becomes a Superfund site[26] is considered a contributing factor to the problem at that site. From a legal perspective, there are arguments both in favor of and against such a decision.[27]

The argument in favor of liability of metal recoverers for environmental problems is best illustrated in the case of a recoverer sending metal concentrate to a smelter for processing. By supplying the concentrates, the suppliers effectively "arranged for disposal or treatment" of hazardous substances as those terms are used in section 107(a)(3) of CERCLA. Also supporting liability is the fact that the supplier is generally aware that the concentrates or ores are treated or processed by the purchaser in a manner that produces waste products.[28] Indeed, depending upon the nature of the industry in question, the concentrate or ore contracts themselves may mention or take into account that waste by-products or impurities will result from the treatment process.

In addition, "tolling" contracts, which are common in some segments of the metals industry, may provide an additional basis of liability for concentrate suppliers. Under these contracts, concentrates are supplied to a smelter, which "tolls"—that is, processes—the ore concentrates, returning the processed metal to the supplier. In an analogous situation, the EPA sought to recover response costs from pesticide manufacturers that hired a bankrupt formulator to formulate their pesticides into commercial-grade products. The manufacturers allegedly owned the technical grade pesticide, the work in process, and the resulting commercial-grade product. They, therefore, retained ownership of the pesticide throughout the formulation and packaging process.[29] The manufacturers attempted to escape liability by claiming that they had no authority or control over the formulator's operations and therefore did not arrange for the "disposal" of hazardous substances. The court rejected this argument. It stressed their ownership of the pesticides and the fact that the generation of pesticide-containing wastes is inherent in the pesticide formulation process. "Any other decision," the court noted, "would allow defendants to simply 'close their eyes' to the method of disposal of the hazardous substances."[30]

Obviously, there are also arguments against recoverer liability. For example, several cases held that the sale of a useful product containing a contaminant does not by itself subject the seller to CERCLA liability.[31] Mr. Ray Cardenas of Encycle/Texas, Inc. agrees, stating that a metal recovery operation which supplies concentrates to smelters is not a generator since it is supplying a substitute for a raw material which is often more pure than the raw material itself.[32]

Presently, uncertainties surrounding the ultimate resolution of regulatory questions add considerable risk to engaging in metal recovery. As more and more Superfund sites are discovered and litigated, resolution of some of these uncertainties may occur.

Government regulation of disposal facilities has caused generators to consider a broad range of metal recovery alternatives for managing their metal-bearing waste. For example, a generator could establish an "in-house" recovery process to treat waste as it comes out of the main production process, benefiting from the recovered metals and the avoided costs associated with disposal. The cost of developing and operating such a process should be compared to the benefits that the generator receives. Alternatively, a generator might send the waste to an independently run recovery facility and pay a fee per ton of waste recovered. Examples of such arrangements might include:

- a government-run and possibly subsidized facility
- a private enterprise offering treatment service of specific waste stream(s) at its own plant
- a contract agreement where an independent company treats the waste generated on the generator's site
- hybrid scenarios combining the above approaches

In all of these cases, a generator should compare the costs of metal recovery to the costs associated with other forms of treatment.

Regulations, then, have helped to make metal recovery an economic alternative to disposal and, at the same time, have defined the market for metal recovery. Figure 9.1 illustrates the options available to today's generator for managing metal-bearing wastes and the role that regulations play in each alternative. Having provided a brief outline of the metal recovery options that are available to a generator, we can now identify some key factors that are important to the eco-

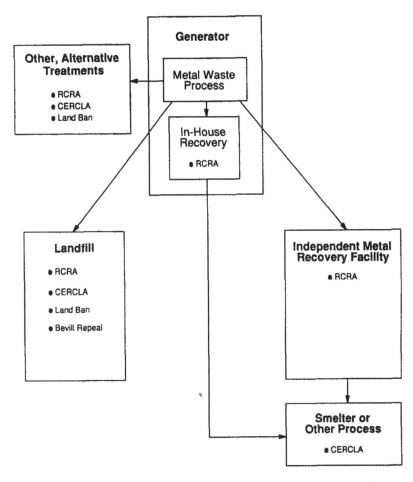

Figure 1. Options for treatment of metal-bearing wastes and their regulatory implications

nomics of metal recovery. These factors are examined in the next section, principally from the viewpoint of the private enterprise. In Section 9.4 we return to examine the impact of the most important of these factors from five unique viewpoints.

9.3 KEY ECONOMIC FACTORS

Based on conversations with operators of metal recovery facilities,[33] the EPA, the BOM, and independent researchers, and a review of the economics of metal recovery literature, we have identified a number of factors that impact metal recovery economics:

A. Waste stream factors
    1. Metal content of waste stream(s)
    2. Volume of waste stream(s) available
B. Operating factors
    1. Operating revenues
    2. Operating costs
C. Plant factors
    1. Capital costs
    2. Capacity
    3. Location
D. Intangibles

These factors are discussed below.

A. Waste Stream Factors

1. Metal Content of the Waste Stream(s)

An important source of revenue for the metal recoverer is the sale of recovered product. The metal content of a waste stream plays a crucial role in maximizing revenues; a high metal-content waste stream can generate more recovered product for a given input of waste than a lower metal-content waste stream. While a high percentage of valuable metals is desirable, there are limiting factors. For example, the relative efficiency with which the metals are recovered is a significant factor in determining the operating cost per unit of metal recovered and has an effect on profitability. The market dynamics (e.g., supply and demand, purity requirements, uneasiness of metal buyers in purchasing recycled products as raw materials) acting on the recovered product are also important.

The number of metals and other chemicals in a waste stream can affect the efficiency of the recovery process. Our investigations uncovered few recycling processes on the market or in pilot testing that are indifferent to the number of metals present in a waste stream. Most metal recoverers would prefer a monometal solution or sludge. Many systems are not designed to accept wastes that contain certain metals or other additives (e.g., cyanide, arsenic, solvents). With the aid of bench- and pilot-scale testing, a metal recoverer must identify the most suitable technology for the given waste stream(s) and design the operation accordingly.

The prospective market for a recovered metal product is metal specific and prices often fluctuate. Metals such as aluminum and iron that are heavily recycled in scrap form (a cheaper means of recycling) are usually poor candidates for metal recovery other than as by-products. High value metals are readily available as scrap, but are also common in a number of aqueous solutions and mixed solid waste streams. These types of metals represent the best economic opportunity to recover metals because of their high selling price per unit. The high recycling rates of precious metals (platinum, gold, and silver) illustrate this phenomenon (see Table 9.4). Due to the relatively high value of most nonferrous metals, waste streams bearing these metals are good candidates for recycling. Metals, other than precious metals, that are most often considered for recovery in working or pilot plant scale include zinc, copper, nickel, chromium, cobalt, cadmium, and vanadium.

2. Volume of Waste Stream(s) Available

In order for a metal recovery endeavor to be profitable, a consistent supply of the appropriate waste stream(s) must be found. Depending on the actual goals of the recovery operation, this could mean centrally locating a recovery plant in an area in which a significant amount of the appropriate waste stream is generated (and alternative disposal is limited) or establishing the recovery operation as an integral part of a

Table 9.4 Major Recycled Metals

Metal	Commodity price 20 February 1991 (dollars/lb.)	Scrap market share[b] (%)	Qualitative ranking of scrap market share[c]	Comments[c]
Platinum	$6324.8	60	Very high	• Scrap market share understates recycling because of increased platinum demand from catalytic converter production. Recycling of converters will lag behind production in the short term
Gold	$5829.6	100	Very high	• Jewelry and industrial uses experience high rates of recycling • Full recycling is probably not really achieved, but is due to inventory accounting procedures
Silver	$59.18	48	Medium to high	• Not highly recycled because of use in products such as silver plate or photographic products which are not extensively recycled
Nickel	$4.83	27	Medium	• Scrap market share understates recycling because of lack of accurate accounting for all scrap. • Real value is probably closer to 50%
Copper	$1.36	60	Medium	• Level of recycling has been stable through most of century
Zinc	$0.65	27	Low	• Low rate overall because of large consumption for destructive end uses such as galvanizing and paints. Also low rate of recycling of zinc in steel flue dusts. (Only one major recycler presently exists: Horsehead Resources)

[a]Commodity prices are from *The Financial Times*, February 21, 1991, and exchange rates are from *The Wall Street Journal*, February 22, 1981

[b]Defined as volume of metal refined from scrap divided by total apparent consumption volume. Figures are taken from Mineral Commodity Summaries 1990, U.S. Bureau of Mines.

[c]From phone interview on February 22, 1991, with Mr. Han Spoel of Shredmet, Inc., St. Hubert, Quebec, Canada.

metal-waste producing plant. The advantage of the central-ized plant is the economy of scale (i.e., costs per unit of input decline as plant size increases) resulting from increased access to higher volumes of suitable waste streams. The advantage of the site-specific plant is the increased techno-logical efficiency derived from a waste stream-specific process.

Supply information, in the form of waste generation vol-ume data, is available from the EPA and in capacity assur-ance plans generated by state governments. Using these reports is helpful in deciding where to position a plant so that it is centrally located to a large supply of the appropriate waste stream(s). For example, in 1987, generators in the state of Illinois generated approximately 647,780 tons of metal-bearing waste,[34] but recovered only 15,610 tons (or 2.4%).[35] The rest of the metal-bearing waste was either landfilled (97,785 tons or 15.1%) or was treated in some other fashion (534,385 tons or 82.5%), mostly by deep-well injection.[36] Considering only the volumes of metal-bearing waste land-filled and recovered (a conservative assumption), the total market for metal recovery is 113,395 tons, of which only 14% is presently being recovered. Illinois clearly has a legitimate need for a metal recovery facility. In fact, Recontek of San Diego, California began operating such a facility in central Illinois in November 1990.

Examining the import/export waste volumes and waste treatment capacities[37] of the state in question and the sur-rounding states is often useful. A state that is a waste exporter is probably short on supply of disposal capacity and might be under political pressure from neighbors to begin carrying its weight. Recycling offers a chance to decrease the state's net exports (and associated transportation costs) and is frequently a more popular alternative to the general public than opening a landfill or other disposal facility. The eco-nomics are further improved if a state is surrounded by a number of low-disposal capacity states requiring generators to transport waste even further for disposal. In the case of Illinois, the surrounding states (Indiana, Michigan, Minnesota, Ohio, and Wisconsin), with the exception of Indiana, all project insufficient capacities for metal recovery by 2009 A.D. (see Table 9.5). These states also project low capacities for stabilization or landfilling, creating an incentive for metal-bearing waste generators in those states to search for means to treat wastes elsewhere. Illinois, as a centrally located state in this region, is an attractive location to capture these exports at least in the short term, since neighboring states do not currently have adequate metal recovery facilities (see projections for 1989 in Table 9.5). Alternately, one might want to locate a facility in Wisconsin because of its low capacity for both metal recovery and landfilling alike.

A metal recovery facility needs a consistent supply of its chosen waste streams. Identifying states or regions with low treatment capacities and high production volumes of these wastes allows a recoverer to improve his chances for accessing an adequate supply of appropriate waste streams.

B. Operating Factors

1. Operating Revenues

Both revenue sources of the recovery operation (fees charged to generators to treat their waste and proceeds from the sale of recovered metals[38]) are subject to market forces and are possible to estimate, at least at any given point in time. Transportation costs are also relevant to the fees that can be charged for treatment.

The maximum fee charged to a generator for recycling waste is, all else being equal, equivalent to the sum of the cost of the least expensive alternative treatment methodology plus the contingent liability that is assumed when utilizing the competing treatment. RCRA now requires stabilization prior to landfilling of many waste metals. If the current rate is $375 per ton[39] for stabilization plus landfill, then a recovery operation, in order to be competitive, could charge a fee equal to or less than the sum of $375 per ton and a contingent liability premium of $38 per ton (as calculated in

Table 9.5 Hazardous Waste Treatment Capacity of EPA Region V (tons)

Treatment	Illinois	Indiana	Michigan	Minnesota	Ohio	Wisconsin	Region V
				1989			
Metals recovery	79,605	84,970	(90)	(2,953)	(2,445)	(420)	158,667
Stabilization	40,506	92,485	822,752	(2,521)	1,282	(1,770)	952,734
Landfill	4,450,342	705,381	1,245,799	(9,903)	2,195,050	(9,855)	8,576,814
Other	663,833	1,515,808	1,099,062	26,153	1,767,531	45,087	5,117,474
Total	5,234,286	2,398,644	3,167,523	10,776	3,961,418	33,042	14,805,689
				2009			
Metals recovery	125,436	83,517	(59)	3,924	(31,692)	1,288	182,414
Stabilization	44,543	82,597	868,913	12,097	23,198	(1,784)	1,029,564
Landfill	3,721	(1,569,400)	(910,821)	492,788	748,547	(137,970)	(1,373,136)
Other	912,902	3,090,020	1,462,428	47,142	1,964,137	47,417	328,782
Total	1,086,602	1,686,734	1,421,461	555,951	2,704,190	(91,049)	7,167,624

Source: "Capacity Assurance Plan", State of Illinois, Illinois Environmental Protection Agency, Springfield, 1989, p. 99
Note: Capacity, as used in this table, refers to the maximum residual volumes of hazardous waste that the state can manage by waste treatment method in the given year. Negative values refer to projected deficiencies. Totals for Region V may not add due to rounding and/or adjustments for surplus capacity.

Table 9.3), or $413 per ton. This assumes that the market recognizes that a premium for contingent liability is appropriate. At a minimum, the metal recoverer can charge the going rate of $375 per ton.

Proceeds from the sale of metal compounds are related to metal prices and are therefore hard to forecast. Although the pricing of a recovered metal product is not necessarily directly linked to the metals commodity market, recovered metal prices will experience many of the same uncertainties. Pricing schemes exist that can reduce this risk by placing a portion of the burden for the metal price volatility on the generator (e.g., refunding a given percentage of the proceeds to the generator in exchange for a higher treatment fee),[40] but reliance on the sale of the metals for a constant level of revenue is not possible.[41] More elaborate strategies exist, such as hedging in the metals commodity market,[42] but these are beyond the scope of this book.

Transportation, while not providing revenue directly to the recovery operation, can have an effect on both the treatment fees and the proceeds from the sale of recovered metals.[43] In the case of a centrally located metal recoverer, the waste is shipped to the plant, and the recovered metals are shipped to the end user of the metal. Shipping of hazardous wastes is an expensive proposition; costs are estimated at 23 cents per ton-mile,[44] or $46 per ton for a load shipped 200 miles. Certainly, comparison of relative transportation costs between the target market and the competing disposal sites will allow a recoverer to optimize the physical location of the plant. However, Recontek indicated that, at this early stage in the development of the industry, transportation really is not important since generators have so few available disposal alternatives and are accustomed to shipping wastes long distances to TSDFs.[45]

Based on all of these factors, a recoverer can determine a benchmark price against which operating costs are compared to predict profitability. Some waste streams will be profitable and others will not.

2. Operating Costs

A metal recovery plant is subject to many of the same operating expenses as a typical chemical plant. These expenses are clearly dependent on the metal recovery technology utilized. Among the most significant costs are

- utilities
- chemical supplies
- direct labor
- indirect labor—analytic technicians, managers
- regulatory costs—environmental professionals

These costs can vary significantly depending on the metal recovery process. For example, a pyrometallurgical process would require a great deal of energy, leading to high utility costs, whereas a hydrometallurgic process would place a much greater demand on water and chemical supplies. Choosing the most technologically effective and efficient process for the waste stream of interest is of utmost importance. This is achievable through deriving the operating costs as functions of throughput volume.[46] These cost functions are useful for determining the optimal design of a facility.

C. Plant Factors

1. Capital Costs

A metal recovery operation is just like any other venture— it needs capital. Obtaining this capital requires financing, which is largely dependent on the risks of the project and the time it takes before the project comes to fruition. Typically, larger companies and companies with good access to financing are most likely to undertake the risks and time commitment required by metal recovery projects.[47]

Initially, a metal recovery venture requires capital to design and test a process at a bench and/or pilot plant scale. Later, the venture will require capital to obtain the necessary plant and equipment as well as to run the operation until sufficient cash flow is developed. Typically, these up-front costs are very high and exhibit large economies of scale (estimates include $3.2 million in equipment capital alone for a 16,500-ton-per-year facility[48] and $20 million in total start-up capital for a 72,000-ton-per-year facility).[49] A metal recovery project is often at a disadvantage in soliciting the required funds because of the unknown risks that are involved and the time required before the recovery plant is self-sufficient and can begin to pay back its investment.

The regulations governing a recovery facility contribute a great deal of uncertainty to the future revenues and costs of a venture. Because of these risks, a bank or venture capitalist or other lender may demand a high return on its investment. On the other hand, if a project can qualify, tax-exempt industrial development bond (IDB) financing may be a source of long-term capital. Tax-exempt IDBs, which are available in every state subject to both federal and state limitations and issuing procedures, are an avenue of financing that recycling firms should explore for project financing. Congress authorized the overall IDB program and is scheduled to review some forms of IDBs for extension.[50] Where a company cannot obtain adequate financing from federal or state programs, it may be advisable for a metal recovery facility to seek a large corporation as a joint venture partner in order to gain access to the finance markets.

Time is an issue as well. Due to the length of time required for design, bench and pilot testing, siting, obtaining a permit, and building the plant, the amount of time from conception until plant operation could exceed 5 years (see Table 9.6). Certainly, returning operating profits to its owners will take at least that long. Again, a large company that has a long-term investment strategy may be more willing and able to wait for a recovery operation to pay off. In addition, once a company has an initial plant in operation, subsequent plants should come on line more quickly.

Most of the metal recovery companies that we have contacted receive their financing from a large parent corpora-

Table 9.6 Duration of Initial Activities[a] for First Plant

Activity	Duration (years)
Bench and pilot testing	1–2
Obtaining plant permits	1–2
Construction of plant	1–2
Initial operation	~ 0.5
Total	~ 3.5–6.5

[a]From personal interview on December 17, 1990, with Mr. Wayne Rosenbaum, Executive Vice President of Recontek, Inc., San Diego, CA.

tion. They are typically set up as subsidiaries, presumably to limit the liability of the parent corporation (although the benefits of limiting liability through this structure continue to diminish), but the recovery venture still enjoys the favorable financing achievable through the larger parent.

2. Capacity

Once the operating cost functions are determined, they are compared with revenues for given throughputs and recovered product volumes, allowing a designer to choose an optimal throughput volume subject to market constraints (e.g., the total market for the waste stream of interest may result in a less than optimal required throughput volume). At the same time, strategic and financing considerations may affect the final decision regarding the capacity[51] of the plant.

Although a plant can operate optimally at the designed throughput[52] for a given treatment fee, the treatment fee is likely to increase over time as more landfills close and as disposal requirements become tougher for metal-bearing wastes. For this reason, it is prudent to build in extra capacity based on projections for increased demand for recycling that will result as the cost differential between recycling and treatment fees increases. This of course assumes that recycling fees and operating costs remain constant. Construction of a larger facility at the outset is much cheaper per volume of capacity than expansion at a later date. Ultimately, the optimal throughput should approach plant capacity. At the same time, the capital requirements should not burden the operation with debt service that it cannot meet. The capacity of a recycling plant should provide the flexibility for the operation to expand as demand increases while not over-stretching the financial stability of the venture.

3. Location

A generator that adds a metal recovery unit to its sole existing plant cannot easily modify its location, but the final siting of a freestanding metal recovery plant is a very important factor in running a successful metal recovery operation. In addition to considering proximity to waste stream supply, a metal recoverer must realize that the local community plays a large role in the permitting and regulatory process. Wherever possible, when considering possible site alternatives, the metal recoverer should evaluate proximity to markets, communication with the chosen community, and the role of the community in the regulatory process.

A metal recoverer can identify a suitable area for siting a plant by analyzing state waste stream availability (from EPA and state documents; see discussion on capacity assurance plans in Appendix B) and relative transportation distances to competitive treatment facilities. After this is accomplished, a recoverer should turn his/her attention to finding a community in which to place the plant.

The community surrounding a plant has input into the regulatory process through public comment and through pressure on government officials. Given that a recycling operation is linked directly to the environment, one of today's hottest political topics, the community will show a great deal of interest. Open communication with the public regarding the risks and benefits of having a metal recovery operation in the community will quite possibly avoid costly surprises (e.g., finding out some portions of the community are nonsupportive well into the permitting process). Recontek related that smaller towns are ideal for communication, since news tends to be exchanged quickly in that type of environment.[53]

The community can also have a significant effect on the regulatory process. A recycling plant can have a large impact on a small town's economy. With community support, the permitting process can be expedited (to as little as a year, according to one source),[54] thereby decreasing the amount of time necessary to have valuable capital tied up. Careful choice of a community in which to build will pay off by speeding up the permitting process and by allowing for smooth interaction with the public and the political environment.

D. Intangibles

A metal recovery operation can have everything going for it on paper, but fail if it does not have the support of regulatory agencies or the industrial community with which it interacts. Consideration given to these groups will allow a recovery operation to survive as an ongoing entity.

Recyclers as a whole do not have a spotless record with regulatory agencies. Rather, there have been and continue to be a number of "sham" recyclers who operate under the name of recycling, but do not, in fact, sell their recovered product. Instead, they profit from the alleged treatment of hazardous waste when they are really just stockpiling the waste in a different form. The existence of this type of activity was one of the driving forces behind RCRA: to ensure that companies treating hazardous waste do so in a legally responsible manner.

Many sham recyclers have closed down as a result of RCRA and, in fact, now comprise a significant number of Superfund sites. For this reason, recycling does not have a positive reputation. Generators saddled with Superfund liability as a result of using a so-called recycler are very hesitant to make the same mistake again. A legitimate metal recoverer has the challenge of selling the generator on the viability of the operation and the product that will result.

Similarly, a metal recoverer must develop a reputation with government agencies of adhering to the law and reacting quickly to agency requests. The metal recoverer who develops a strong rapport with both the industrial community and the regulatory agencies helps to ensure that the only challenges a recovery operation faces are operational in nature.

In summary, the above issues cannot be considered individually. Rather, they all combine to allow a prospective metal recoverer to evaluate potential waste streams, choose the most economically efficient process, scale a plant to optimize the profits of the venture, and manage the entire procedure so that the government and the public are satisfied—not a small task.

9.4 VARIOUS METAL RECOVERY SCENARIOS

Earlier in the chapter, we identified several different perspectives from which to view metal recovery. Each perspective, or scenario, implies a distinct set of goals, thus changing how the factors identified above relate to each other. The scenarios and their accompanying goals are summarized in Table 9.7. For each of these scenarios, we will expand on the goals and illustrate the relevant economic factors.

A. Government-Run Facility

A state or the federal government could finance and operate a metal recovery facility. The goals of such an operation might include:

- *Decrease the volume of waste entering the nation's landfills*— RCRA was passed to minimize the amount of hazardous waste entering the dwindling capacity of the nation's landfills. A government-sponsored metal recovery plant could provide an alternative means to landfilling or stabilizing plus landfilling wastes.

Table 9.7 Metal Recovery Scenarios

	Scenario	Goal
1.	Government-run or contracted, state/multi-state facility	Reduce highest volume of waste entering landfills and protect strategic reserve of metals in the U.S.
2.	Private enterprise offering treatment service of specific waste stream(s) at its own plant location	Profit from treatment fees and sale of metals recovered
3.	Generator of metal-bearing wastes	Minimize disposal fees, reduce contingent liability, and recover metals for sale or use in process
4.	Contract treatment operation that operates on generator's facility	Profit from fees paid to manage facility which reduces contingent liability and recover metals for resale
5.	Hybrid of above scenarios	Situation-dependent

- *Reduce the nation's dependence on foreign supplies of strategic metals*—the government could also provide metal recovery for waste streams containing strategic metals. Recovery of these metals inside the United States could significantly decrease dependence on imports.
- *Reduce the cost of treatment at Superfund cleanup sites*—by offering a cheaper alternative to stabilization and landfilling as a treatment of hazardous wastes, the government can reduce the cleanup costs at a number of future Superfund sites.

The factors that play an important role in achieving these goals are as follows:

- *Metal content of waste stream(s)*—in the interest of accepting the largest volume of waste possible, the government would treat some waste streams with low percentages of metals, decreasing the revenues (on a per unit basis) available from the sale of recovered product.
- *Operating costs*—similarly, accepting large, inconsistent volumes and a mix of waste streams would mean higher operating costs due to the inefficiencies that develop when vari-

ety and inconsistency are introduced into a chemical process.
- *Location*—capturing the greatest waste volume dictates a plant location that is closer than alternative disposal options. This may lead to an increase in land, utility, and other costs if the plant is situated close to generators in an industrial/urban setting.

In summary, a government-sponsored recovery plant could be beneficial, but to fully serve its purposes, extensive subsidization from the government is probably required.

B. Private Enterprise Offering Treatment Service of Specific Waste Stream(s) at Its Own Plant

A private enterprise could, separately from any other enterprise, develop a process to treat select waste streams and build and operate a treatment facility which utilizes the specific process. The goals of such a venture include:

- *Generating maximum profits*—a private venture is interested in maximizing return on investment for the owners.
- *Gaining market share*—for the sake of long-term viability, a private metal recovery operation should build market share (i.e., develop a large base of waste stream suppliers) to ensure a lasting and reliable supply of the required waste streams.
- *Avoiding environmental liability*—environmental liabilities due to the release of hazardous materials are expensive to correct and threaten the regulatory acceptance of the venture.
- *Promoting worker safety*—similar to environmental liabilities, worker injuries can result in expensive corrective measures and additional scrutiny from regulatory agencies.
- *Gaining technical expertise*—a private enterprise is dependent on its technology to improve profits and identify and develop new opportunities.

Factors which play an integral role in achieving these goals include:

- *Metal content of waste stream(s)*—high metal content generates higher revenues per unit of waste treated, contributing to higher profits.
- *Volume of waste stream(s) available*—a private venture is very dependent on the supply of its chosen waste stream(s). Inconsistency or interruption of supply can lead to inefficiencies and thus greater operating costs.
- *Operating costs*—by choosing a specific waste stream(s), a metal recoverer can custom fit the process so as to minimize operating costs.
- *Capital costs*—a private metal recoverer has the challenge of finding financing for the design, construction, and initial operation of the recovery plant. Financing can significantly affect the cash flow and ultimate profitability of a metal recovery project. For this reason, a small venture should seek out a large corporate partner with some degree of leverage in the financial markets.
- *Other regulatory issues*:
  - *Regulation of waste stream(s)*—recovering metals from a heavily regulated waste stream is desirable, since alternative treatment fees are high. This allows for higher recycling fees, which lead to higher profits.
  - *Regulation of the recovery operation*—to minimize environmental and worker liability, a metal recoverer must be familiar and comply with all relevant government regulations. This requires a recoverer to hire environmental professionals (e.g., engineers, lawyers, and consultants).

As can be seen, a private venture has a great deal of flexibility towards accomplishing its goals from an operational sense, but is largely dependent on regulations (both of waste streams and the metal recovery facility) as the ultimate factor in determining profits.

C. Generator of Metal-Bearing Waste(s)

A generator of metal-bearing wastes could construct a metal recovery facility as an integral process within a larger plant, treating the wastes as the main process produces them and feeding the recovered metals as a raw material back into the plant. The goals of such a "closed-loop" approach include:

- *Improving the profitability of the main plant*—this can be accomplished by decreasing or eliminating the disposal fees and the associated contingent liabilities that the plant faces when sending wastes off-site. The recovered metals also reduce raw material costs.
- *Exploiting economies of scale*—a generator should seek to spread overhead over value-adding processes as much as possible.
- *Exploiting process synergies*—a plant probably generates a consistent and reliable waste stream. By custom-fitting a process to that waste stream, a generator can help ensure cost-effective metal recovery.

Whether or not a generator can accomplish these goals depends largely on the following factors:

- *Metal content of waste stream(s)*—high percentages of metal in a waste stream would lead to a greater reduction in the cost of raw materials that a plant experiences.
- *Volume of waste stream(s) available*—a generator's waste volume is determined solely by production in the main plant. As a result, the volumes of waste that are produced for the recovery process are probably less than optimal. This could hurt the overall economic feasibility of the project.
- *Operating costs*—an existing plant already has an infrastructure in place for dealing with environmental and technical overhead. By spreading this overhead over additional operations, a generator can decrease the overhead costs per unit.
- *Other regulatory issues*:
  - *Regulation of the waste stream(s)*—a highly regulated waste stream would cost more to treat off-site, so a generator with such a waste stream has a greater incentive to develop metal recovery capabilities.
  - *Regulation of the recovery operation*—a metal recovery operation could require a RCRA permit, increasing government and regulatory scrutiny of all operations at the facility.

A metal recovery operation on a generator's plant site can provide economic savings only if any additional costs due to suboptimal throughput volumes of waste are counterbalanced by the savings from process synergies and economies of scale. Additional regulatory considerations for the entire facility also complicate the decision.

D. Contract Treatment Operation on a Generator's Facility

This type of operation is simply a variation of the last scenario, except that the generator does not have either the in-house knowledge to design and run a metal recovery operation or the desire to bear the financial risks of operating the plant on a day-to-day basis. In either case, the generator can hire someone to design and operate a facility on a contract basis.

E. Hybrid

The above scenarios represent the most obvious situations where metal recovery would make sense. Many other possibilities exist, such as:

- joint ventures between generators and treatment firms
- government contracts for specific waste stream treatments
- recovery facilities as a component of a smelter's operations

In any set of circumstances, the economic factors we have discussed will combine to determine the feasibility and, ultimately, the final design of a metal recovery facility.

9.5 POSSIBLE HELP FROM THE GOVERNMENT

In addition to the economic factors outlined above, the responsiveness of the federal government to recycling as an industry can have a substantial effect on metal recovery. For example, there are in existence federal economic development programs (e.g., loans and bond guarantees from the Small Business and the Farmers Home and Economic Development Administrations) which assist small-business metal recoverers. Federal tax laws also favor pollution control through investment credits. The addition of other government programs would help make metal recovery even more economically feasible. Programs that provide metal recoverers with regulatory or tax advantages are most helpful.

One metal recoverer, Encycle/Texas, Inc., has proposed that the government take the following actions:[55]

- require recycling as the treatment for a waste which can be recycled (i.e., make recycling the best developed alternative treatment or BDAT where possible
- increase taxes on land disposal facilities, focusing on wastes which can be recycled
- develop special permits for recyclers to assure generators and the public that recycling is the preferred method of waste management
- exempt transporters from state and federal taxes for hauling waste to and from recycling facilities
- give preference to recycled products in federal procurement programs
- give preference to recyclers in the permitting process at the state and federal levels
- make additional tax savings available to recyclers, such as investment tax credits

In addition, Encycle has promoted the idea that metal recovery could be used as a treatment method for metal waste from Superfund sites.[56]

The states could also play an important role in promoting metal recovery. They could offer tax-exempt financing, loans, and mortgage insurance to small contractors and manufacturers who operate facilities that reduce, recycle, or process hazardous wastes. States can also offer technical assistance and matching grant programs to assist the evolution of waste metal recovery.

One could also envision the small generator or recoverer becoming dependent on outside consultants for technical expertise because of the large investments required for full-time professionals. In fact, there have been a number of studies regarding disposal and recycling costs for the metal finishing industries, largely under federal (EPA) funding.[57,58] The government could expand its program.

Clearly, the Congress, the EPA, and state governments must be convinced that metal recovery is both safe and worthwhile before such measures will be taken. If these suggestions are accomplished, a greater volume of metal waste could be economically managed through metal recovery.

9.6 THE DECISION

This chapter has highlighted the key factors that one should consider when evaluating the economic feasibility of a metal recovery project. These factors represent the most important considerations on a strictly qualitative level. Little discussion on quantifying these factors has taken place, although this represents the next step that evaluators should complete before actually proceeding with a metal recovery project. Through sensitivity analysis, an economic model can determine the most important factors and quantify the risks that these factors bring to the overall success of the project. The only way to perform such an analysis is on a case-by-case basis, and only when all issues, regulatory and economic, are understood. Additionally, one should not forget management techniques that could ensure the success of such a venture. These techniques might include:

- pricing schemes to minimize the metals price risks
- securing preferred government financing or grants for seed and/or working capital
- investing in and exploiting recovery technology, even at a loss, for future profitable application

In summary, when considering a metal recovery project, one should specifically outline one's goals and consider all of the key factors outlined previously. At this point, a preliminary assessment of the feasibility of a metal recovery plant can be made. Assuming that the qualitative assessment is positive, the next step is to perform an in-depth economic analysis, similar to a business plan that factors in the elements identified as unique to this industry, making full use of modeling and sensitivity analysis. This will allow the evaluation of risks, opportunities, and management considerations of the project.

As regulation of metal-bearing wastes increases and landfill disposal dwindles, the results of these types of analysis are likely to show that metal recovery projects are economical and applicable to a broadening spectrum of metals.

NOTES

1. This chapter was contributed by Philip L. Brooks, Director of the Litigation Consulting practice of Arthur Andersen & Co.'s Pittsburgh, PA office, and George Hansen, Associate, and Laurel A. McCarthy, Research Assistant, of the management and economic consulting firm of Putnam, Hayes & Bartlett, Inc. The views expressed here are the authors' and do not necessarily represent those of either firm.
2. "1987 National Report of Hazardous Waste Generation and Treatment, Storage and Disposal Facilities Regulated Under RCRA, U.S. Environmental Protection Agency (forthcoming)." This information, which reports 1986 data, was provided by Mr. Michael Burns of the U.S. Environmental Protection Agency.
3. F006 is the EPA alphanumerical classification for wastewater treatment sludge from electroplating operations. (Source: 40 *Code of Federal Regulations* (CFR) *Part 264*).
4. "1987 National Report of Hazardous Waste Generation and Treatment, Storage and Disposal Facilities Regu-

lated Under RCRA (forthcoming)." This information, which reports 1986 data, was provided by Mr. Michael Burns of the U.S. Environmental Protection Agency.

5. "The Nation's Hazardous Waste Management Program at a Crossroads: The RCRA Implementation Study," U.S. Environmental Protection Agency, Office of Solid Waste and Emergency Response (July 1990), p. 7.

6. "EPA Repeals Bevill Exclusion for 17 Mining Waste Streams," *Hazardous Waste News*, Vol. 11, No. 35 (August 28, 1989).

7. "RCRA's Solid Waste Regulation and Its Impact on Resource Recovery in the Minerals Industry," Shaun D. Peterson, U.S. Bureau of Mines (September 1990), p. 3.

8. "EPA Repeals Bevill Exclusion for 17 Mining Waste Streams," *Hazardous Waste News*.

9. Ibid., 11(35)(1989).

10. See: "Land Disposal Inventory for First Third Scheduled Wastes," Federal Register (F.R.), V. 53, No. 159, August 17, 1988; "Land Disposal Restrictions for Second Third Scheduled Wastes," F.R., V. 54, No. 120, June 23, 1989; "Land Disposal Requirements for Third Third Scheduled Wastes," F.R., V. 54, No. 224, November 22, 1989.

11. "The Nation's Hazardous Waste Management Program at a Crossroads; The RCRA Implementation Study," p. 43. Percent calculated by dividing number of facilities closing (1,273) by total number of facilities (1,273 + 194) times 100.

12. Tipping fee refers to the charge levied on the generator for each ton of waste that its transporter delivers to the landfill. It does not include other potential disposal costs that may be incurred, such as stabilization or taxes.

13. "Review of Activities of Firms in the Commercial Hazardous Waste Management Industry: 1983 Update," Booz, Allen & Hamilton Inc. for the U.S. Environmental Protection Agency, Office of Policy Analysis (November 30, 1984), Exhibit III-6.

14. "1986–1987 Survey of Selected Firms in the Commercial Hazardous Waste Management Industry," U.S. Environ-

mental Protection Agency, Office of Policy Analysis (March 31, 1988), Exhibit 3-4.

15. The NPL is a list of sites slated for further investigation by the EPA as required by the National Contingency Plan under SARA. Most NPL sites become Superfund sites. (Source: *40 CFR Part 300*, p. 5.)

16. "Methods of Site Remediation," *Poll. Eng.*, Vol. 22 (November 1990), p. 58.

17. *40 CFR Part 300*, p. 23.

18. From phone interview on January 18, 1991 with Mr. Ray Cardenas, President of Encycle/Texas, Inc., Corpus Christi, TX.

19. From phone interview on February 22, 1991 with Mr. Philip Edelstein, Director of CP Chemicals, Environmental Recovery Services (ERS), Ft. Lee, NJ.

20. "RCRA's Solid Waste Regulation and Its Impact on Resource Recovery in the Minerals Industry," p. 2.

21. Ibid.

22. "Recyclable Minerals: The Cost of Regulation," Shaun D. Peterson, *Minerals Today* (January 1991), p. 21.

23. Ibid., p. 22.

24. This activity is fairly low on the agenda and, therefore, may not be addressed in 1991.

25. A sham recycler is a facility that stores and/or treats hazardous waste under the guise of recycling. Many such facilities have been named as Superfund sites.

26. Smelters, potential customers for metal recoverers, have been named as Superfund sites. See: Superfund sites: Palmerton Zinc (EPA/ROD/RO3-871036, September 1987) and Gould site (EPA Record of Decision, March 1988).

27. From correspondence dated January 15, 1991, with Richard Mancino, Esquire, of Willkie Farr & Gallagher, New York, NY.

28. For example, see *United States v. Aceto Agricultural Chemicals Corp.*, 872 F.2d 1373 (8th Cir. 1989) (defendants were aware that wastes were generated during the manufacturing process); and *New York v. General Electric Co.*, 592

F. Suppl. 291, 297 (N.D.N.Y. 1984) (GE sold used trans-
former oil containing PCBs to a drag strip, which used
the oil for dust control; the court denied GE's motion to
dismiss because GE allegedly arranged for the drag strip
to take its used oil "with knowledge or imputed knowledge" that the oil would be deposited on the land).

29. *United States v. Aceto Agricultural Chemicals Corp.*, p.
1378.
30. Ibid, p. 1382.
31. See, for example, *Florida Power & Light Co. v. Allis-
Chalmers Corp.*, 893 F.2d 1313 (11th Cir. 1990) (FP&L purchased transformers containing PCBs from Allis-
Chalmers, used them for 40 years, and then made the
decision to dispose of them on-site; the court held Allis-
Chalmers not liable under CERCLA).
32. From phone interview on January 18, 1991, with Mr. Ray
Cardenas, president of Encycle/Texas, Inc., Corpus
Christi, TX.
33. We are thankful to the following metal recovery companies for their input to this section: CP Chemicals, Environmental Recovery Services (ERS), Ft. Lee, NJ; Encycle/
Texas, Inc., Corpus Christi, TX; and Recontek, Inc., San
Diego, CA.
34. Total hazardous wastes generated in Illinois in 1987 com-
prised 1,891,629 tons.
35. "Capacity Assurance Plan, State of Illinois," Illinois Environmental Protection Agency, Division of Land Pollution Control (October 17, 1989), Tables 1, 2-1, 4-1.
36. Ibid.
37. Capacity refers to the maximum volume of hazardous
waste that a state can manage with a given treatment
methodology over a specified time period.
38. The terms "fee" and "proceed" do not simply apply to a
private metal recoverer. They could also apply as transfer prices in the case of operating a recovery plant as part
of a generator's metal-bearing waste production facility.
39. Estimated from the stabilization cost and volume data
from "Regulatory Impact Analysis of the Land Disposal

Restrictions for Third Third Scheduled Waste Proposed Rule," ICF for the U.S. EPA, Washington, DC (November 15, 1989), Exhibits 3-1 and 3-6. This cost does not include transportation expenses.

40. This is the method Recontek, Inc. employs.

41. Some recyclers engage in refinement of the metals that are recovered. Where such value-added activity takes place, metal recycling provides a less expensive source of raw materials. Under this circumstance, the discounted cost of raw materials helps improve the profitability of the operation.

42. See, for example, R. A. Brealey, and S. C. Myers, *Principles of Corporate Finance* (New York: McGraw-Hill Book Co., 1988), chap. 25.

43. In either of these cases, transportation matters regardless of who pays the transportation fee. In the case of treatment, a metal recoverer who pays for transportation would just raise fees accordingly, whereas a generator who pays for transportation costs would just consider the cost on top of the base fee. In either situation, transportation adds equally to the cost paid by the generator.

44. "1986–1987 Survey of Selected Firms in the Commercial Hazardous Waste Management Industry," U.S. EPA, Exhibit 3-4. This figure represents a 1987 cost.

45. From personal interview on December 17, 1990, with Mr. Wayne Rosenbaum, Executive Vice-President of Recontek, Inc., San Diego, CA.

46. Throughput volume refers to the volume of waste that enters the system and is processed over a given period of time. The term should not be confused with the volume of metal product that is recovered from the recycling process.

47. A note here about financing: Traditional finance theory would state that investment and finance decisions are separate, making the type of financing or even the debt rate immaterial in evaluating a metal recovery project. We maintain that these issues affect tax shields and cash flows which are vital to a single-project firm. (For more

background see, for example, R. A. Brealey and S. C. Myers, *Principles of Corporate Finance*, 3rd ed. [New York: McGraw-Hill, 1988, chap. 19.]) For this reason, we consider project financing a contributing factor in determining the economic attractiveness of a metal recovery project.

48. "Economic Feasibility of a State-Wide Hydrometallurgical Recovery Facility," Ray O. Ball et al., pp. VIII-9, VIII-10 and VIII-12 (in *Metals Speciation, Separation and Recovery*, J. Petterson and R. Passino, Eds.; see Bibliography). We have estimated capacity from this document by multiplying operating days (260) by volume per day (105,000 lbs/day).

49. From personal interview on December 17, 1990 with Mr. Wayne Rosenbaum, Executive Vice-President of Recontek, Inc., San Diego, CA.

50. From phone interview on February 25, 1991 with Mr. Kevin McCarty of the Council of Industrial Development Bond Issuers, Washington, DC.

51. Capacity refers to the total volume of waste that can be treated by a plant.

52. Optimal throughput does not have to match a plant's capacity. For example, a plant with a capacity of 5000 tons per month may operate optimally at 80% of capacity, or with a throughput of 4000 tons per month, depending on market conditions at the time.

53. From personal interview on December 17, 1990, with Mr. Wayne Rosenbaum, Executive Vice-President of Recontek, Inc., San Diego, CA.

54. Ibid.

55. "Metals Reclaimer Urges Agency to Put RCRA on Track," *Hazmat World* (November 1990), p. 50.

56. Ibid.

57. Stinson, M. K. "Group Treatment of Multicompany Plating Waste, Tontin, MA Silver Project," EPA-600/2-710-102; PB 80-102825 (1979).

58. "Control and Treatment Technology for the Metal Finishing Industry: In-Plant Changes," EPA–625/8-82-008 (1982).

BIBLIOGRAPHY

Authors Available

Agoos, A. "Metals Reclaimers Find Slim Pickings," *Chem. Week* (May 14, 1986), pp. 21–23.

Ball, R. O. et al. "Economic Feasibility of a State-Wide Hydro-metallurgical Recovery Facility," *Metals Speciation Separation and Recovery*, J. Petterson and R. Passino, Eds. (Chelsea, MI: Lewis Publishers, Inc., 1987).

Basta, N. "Metals Recyclers Warily Eye New Sources," *Chem. Eng.* (December 1990), pp. 29–35.

Borner, A. J. and B. Perry. "Metals Reclaimer Urges Agency to Put RCRA on Track," *Hazmat World* (November 1990), pp. 48–51.

Bower, B. T., Ed. "Resources for the Future," *Regional Residuals Environmental Quality Management Modeling* (Washington, DC, 1977).

Brooks, C. S. "Marketing of Recycled Waste Metals," paper presented to Recycling Task Force, Connecticut Hazard Waste Management Service, June 2, 1986.

Brooks, C. S. "Metal Recovery from Industrial Wastes," *J. Metals* 38(7):50–57 (1986).

Brooks, C. S. "Sources for Marketing Information on Metal Use and Recycling," paper presented to Recycling Task Force, Glastonbury, CT, June 2, 1986.

Brooks, P. L. and T. Slevin. "Strategic Metal Recovery, Business Plan," Carnegie Mellon University, Pittsburgh, PA (1982).

Cammarota, V. A. Jr. Personal communication (1982).

Coleman, R. T. et al. "Sources and Treatment of Wastewater in the Nonferrous Metals Industry," Indiana Environmental Research Lab, EPA-600/2-80-074 (1980).

Guweh, S. "Adding Financial Incentives," *Approaches to Source Reduction* (Berkeley, CA: Environmental Defense Fund, June 1986), pp. 115–132.

Hauck, J. and S. Masoomian. "Alternate Technologies for Wastewater Treatment," *Poll. Eng.* (May 1990), pp. 81–84.

Howe, C. W. *Natural Resource Economics, Issues, Analysis, and Policy* (New York: John Wiley & Sons, 1979), pp. 140–144; 203–208.

Islam, S. "Recycling Industry Has Taken a Stand: Will Fight 'Ill-Conceived' Legislation," *Am. Met. Mark.* (November 2, 1990), p. 8.

Krieger, J. "Hazardous Waste Management Database Starts to Take Shape," *Chem. Eng. News* (February 6, 1989), pp. 19–21.

Kuster, T. "Portable Recovery Plants Foreseen," *Am. Metal Mark.* (February 19, 1990), p. 7.

Leff, D. K. "Economic Development Programs to Assist Hazardous Waste Generators," Connecticut General Assembly (1985).

London, P. F. "EEC's Aim: Improve Recycling Methods," *Am. Metal Mark.* (December 8, 1989), p. 9.

Meschter, E. "Zia Constructing Plant for Processing EF Dust," *Am. Metal Mark.* (June 6, 1990), p. 9.

Pahlman, J. E. and S. E. Khalafalla. "Use of Lignochemicals and Humic Acids to Remove Heavy Metals from Waste Streams," Bureau of Mines, U.S. Department of the Interior (1988).

Patel, Y. B., M. K. Shah, and P. N. Cheremisinoff. "Methods of Site Remediation," *Poll. Eng.* 22(12): 58–66 (1990).

Peterson, S. D. "RCRA's Solid Waste Regulation and Its Impact on Resource Recovery in the Minerals Industry," Bureau of Mines, U.S. Department of the Interior (1990).

Peterson, S. D. "Recyclable Minerals: The Cost of Regulation," *Miner. Today* (January 1991), p. 21.

Rosenbaum, S. W. "The Economic Case for the Recycling of F006 Wastewater Treatment Sludges from the Plating Industry," Recontek, San Diego, CA (December 1990).

Smith, J. D. "Recontek: A Recycling Alternative for F006 and D002 Wastes." *El Digest* (June 1990).

Spoel, H. "The Current Status of Scrap Metal Recycling," *J. Metals* 42(4):38–41 (1990).

Twidwell, L. G. "Metal Value Recovery from Hydroxide Sludges, Final Report," U.S. Environmental Protection Agency, EPA/600/2–85/128 (1985).

Twidwell, L. G. and D. R. Dahnke. "Metal Value Recovery from Metal Hydroxide Sludges: Removal of Iron and Recovery of Chromium," U.S. Environmental Protection Agency, Cincinnati, Ohio, EPA/600/2-88/019 (1988).

Viani, L. "Ground Broken at High-Tech Battery Recycling Plant," *Am. Metal Mark.* (July 20, 1990), p. 4.

Authors Not Available

"112th American Institution of Mechanical Engineers (AIME) Annual Meeting Technical Program with Abstracts," *J. Metals* 34(12) (1982).

"113th American Institution of Mechanical Engineers (AIME) Annual Meeting Technical Program with Abstracts," *J. Metals* 35(12) (1983).

"1985 National Report of Hazardous Waste Generators and Treatment, Storage and Disposal Facilities Regulated Under RCRA," U.S. Department of Commerce, NTIS Publication No. PB89-187652 (1989).

"1986–1987 Survey of Selected Firms in the Commercial Hazardous Waste Management Industry," Office of Policy Analysis, U.S. Environmental Protection Agency (1988).

"1989 Capacity Assurance Plan for the State of Texas," Texas Water Commission, LP89-05, October 1989.

"1990 Extractive & Process Metallurgy Fall Meeting; Recycling of Metals and Engineering Materials," The Minerals, Metals & Materials Society (TMS) (October 28, 1990).

"The 1991 Business Communications Co. (BCC) Environmental Conference Outline," BCC (1991).

"Capacity Assurance Plan, State of Illinois," Illinois Environmental Protection Agency, Division of Land Pollution Control (1989).

"Compliance Recycling Plans Plant in Chicago," *Am. Metal Mark.* (August 21, 1990), p. 7.

"Economic Analysis of Proposed Effluent Standards and Limitations for the Metal Finishing Industry," Booz, Allen & Hamilton Inc., U.S. Environmental Protection Agency, EPA-440/2-82-004 (1982).

"Economic Impact Analysis of RCRA Interim Status Standards," Arthur D. Little, Inc., Cambridge, MA (1981).

"Economic Incentives for the Reduction of Hazardous Wastes," ICF Consulting Associates, Washington, DC (December 18, 1985).

"EPA Repeals Bevill Exclusion for 17 Mining Waste Streams," *Hazardous Waste News* 11(35)(1989).

"Extent of the Hazardous Release Problem and Future Funding Needs: CERCLA Section 301(a)(1)(C) Study," Office of Solid Waste and Emergency Response, U.S. Environmental Protection Agency (1984).

Facts 1989 Yearbook. (Washington, DC: Institute of Scrap Recycling Industries (ISRI), Inc., 1990).

Gould Site (EPA Record of Decision 10, March 1988).

"Heckett Unit, MKS Sign Metal Reclamation Pact," *Am. Metal Mark.* (May 18, 1990), p. 8.

"Incineration and Treatment of Hazardous Waste," U.S. Environmental Protection Agency, EPA/600/9-85-028 (1985).

"Land Disposal Inventory for First Third Scheduled Wastes," *Fed. Reg.* 53(159) (1988).

"Land Disposal Requirements for Third Third Scheduled Wastes," *Fed. Reg.* 54(224) (1989).

"Land Disposal Restrictions for Second Third Scheduled Wastes," *Fed. Reg.* 54(120) (1989).

"Mineral Commodity Summaries 1982," Bureau of Mines, U.S. Department of the Interior (1982).

"Mineral Commodity Summaries 1990," Bureau of Mines, U.S. Department of the Interior (1990).

"Minerals in the U.S. Economy: Ten-Year Supply-Demand Profiles," Bureau of Mines, U.S. Department of the Interior (1979).

"The Nation's Hazardous Waste Management Program at a Crossroads," Office of Solid Waste and Emergency Response, U.S. EPA Report 530-SW-90-069 (1990).

"Nonfuel Mineral Production in the United States," Bureau of Mines, U.S. Department of the Interior (1988).

Northeast Industrial Waste Exchange Listings Catalog, Issue No. 9 (August 1983).

"Palmerton Zinc," U.S. Environmental Protection Agency, EPA/ROD/RO3-871036 (1987).

"Recovered Value from Electroplating Industry Waste," *Metal Finishing* (June 1990), pp. 85–92.

"Regulatory Impact Analysis of the Land Disposal Restrictions for Third Third Scheduled Wastes Proposed Rule," ICF Incorporated, Office of Solid Waste, Economic Analysis Staff, U.S. Environmental Protection Agency, EPA Contract No. 68-01-7290 (1989).

"Review of Activities of Firms in the Commercial Hazardous Waste Management Industry; 1983 Update," Booz, Allen & Hamilton Inc., Office of Policy Analysis, U.S. Environmental Protection Agency (1984).

"Technical Programs with Abstracts," *J. Metals* 32(11) (1980).

"The Minerals, Metals & Materials Society (TMS) Annual Meeting Technical Program with Abstracts," *J. Metals* 39(10) (1987).

U.S. Code of Federal Regulations (CFR), Vol. 40, Part 264 (1990).

U.S. Code of Federal Regulations (CFR), Vol. 40, Part 300 (1990), p. 23.

"Waste Treatment," *Business Opportunity Reports* (1990).

10

Promising Directions

The appropriate audience for this book, as stated in the preface, is

1. industrial managers, engineers and scientists actively involved with the development of metal recovery technology and/or hazardous waste disposal alternatives
2. industrial consultants
3. government personnel concerned with recycling and hazardous waste managers
4. legislative personnel needing knowledge of the availability of technologically feasible separation processes

Educators and environmentalists should also find the book useful for learning about alternatives for recycling and recovery of metals for industrial waste effluents.

In Chapter 1, three incentives are identified for recovering nonferrous metals from industrial wastes. They are:

1. to better meet state and federal emission regulations
2. to reduce disposal costs by minimizing volume and toxicity and generating recovered metal credits
3. finally to enhance the conservation of resources

In addition, a general overview is provided of the amount of the principal commercial nonferrous metals present in current hazardous waste streams and the extent of recycling.

In Chapter 2, the current federal hazardous waste regulations applicable to metals are identified, discussed briefly, and guidance is provided to sources for more comprehensive information.

In Chapter 3, the concept of resource conservation is developed as it applies to waste metal recovery and its relevance to waste management.

10.1 SEPARATION TECHNOLOGY

Chapters 4 through 7 identify the principal separation processes relevant to metal recovery from industrial waste streams. In addition, a considerable effort has been made to survey separation process, both experimental and in practice over the past 10 years. Copper and nickel are emphasized since they are the principal nonferrous metals appearing in significant volume in industrial waste streams and commanding reasonably high prices in the metal recovery markets.

The principal emphasis in Chapters 4 to 7 is on "hydrometallurgical" technology since it offers the most promise for the majority of metal waste streams in regard to their physical condition. There is a great deal of very relevant information on appropriate separation processes in the hydrometallurgy literature[1 to 6] and in recent symposia and publications on separation technology.[7] These sources have their limitations, however, because the conventional treatment in hydrometallurgy is concentrated on mining and extractive metallurgy from ores, which are usually quite different from industrial waste streams. The separation technology literature, on the other hand, is concentrated on the chemical process industries. Based on my 10 years of experience working in this field, I believe that there is a very urgent need for a more focused treatment of the "hydrometallurgy" of industrial waste streams. It might very appropriately be taught as a special aspect of the usual hydrometallurgy curriculum, where currently the emphasis is on extraction of metals from ores. The present book does not presume to be a textbook of waste stream hydrometallurgy, but it is offered as an introduction to the subject and as an identification of the need for an in-depth treatment of this subject area in engineering course work. The present text also provides a brief discussion of the chemistry of the relevant separation processes, but does not address the engineering of the processes or the specifics of the hardware required. Considerable effort was expended, however, to present as comprehensively as possible each of the major chemical and physical separation processes and how each is applied to metal recovery from industrial wastes.

Chapter 8 deals with recent and current efforts to assemble the unit processes of separation technology to successfully cope with two types of mixed-metal industrial wastes: metal finishing industry wastes and spent catalysts.

10.2 ECONOMICS

Chapter 9 identifies and discusses a number of factors which combine to determine the economic feasibility of metal recovery from hazardous waste streams. Regulation of the waste management industry is the single most important factor in determining the economic feasibility of a metal recovery project. Additional factors that drive the feasibility of a metal recovery project include: metal content and volumes of the waste stream(s) available; operating costs and revenues; capital costs; plant capacity and location; and intangibles. The fundamental economics of conventional process design are not dealt with in detail since they are regularly addressed through established quantitative procedures in the chemical process and economics literature. Rather, the unique aspects of metal recovery are brought forth in an effort to show how they may ultimately affect the economic feasibility of a project.

Multiple scenarios exist in which an organization could potentially benefit from the operation of a metal recovery facility. A plant which treats and recovers metals from a number of hazardous waste generators is just one of these scenarios.* Other scenarios include: a generator operating a recovery facility on-site for the purpose of reducing hazardous waste liabilities (both fees and contingent liabilities); a government operation that is interested in decreasing the overall amount of hazardous waste landfilled in a state or multistate area; or a contract operation whereby a company derives revenues by designing, building, and operating recovery plants for various clients. For each of these alternatives, the changes in the economic factors for a given situation are outlined. The most important factors in almost any scenario are regulation, metal content, and the volume of the waste stream that is available. All other factors vary in relative importance, depending on the particular metal recovery operation.

An additional factor that may affect the success of metal recovery projects in the future is the role of the government in promoting metal recovery. The economics chapter identifies a number of steps that one recycler has proposed the government take to promote metal recovery, including tax incentives, financial assistance, mortgage guarantees, technical assistance, and regulatory preference for recovery operations. By implementing any of these steps alone or in combination, the government may stimulate growth in metal recovery by making it more economically feasible.

The economic success of a metal recovery venture is then dependent on a number of things. Fundamental factors for a given situation predict economic success, and these factors can, in turn, be influenced by outside environmental forces such as liquidity of the waste market and/or government incentives. A waste metal recoverer can operate successfully given the correct combination of sound economic fundamentals and a favorable operating environment.

---

*A private venture or existing companies with well-developed client bases in the hazardous waste business, like Chemical Waste Management, Ensco, or Rollins, might try this route. The existing companies could open a plant to supplement their landfill and incineration services.

10.3 CURRENT TRENDS

The framework within which recovery is accomplished includes the generator, the treatment and disposal facility, and the recycle vendor or manufacturer that can utilize recycled metals as raw materials. The most obvious economic advantages accrue to the generator for exercising the recovery options and minimizing his hazardous waste liability exposure. The generator holds control of the production process and is, therefore, able to design operations which minimize waste and recover materials in the most advantageous condition for recycling. For manufacturers in metal extraction and processing industries or chemical process industries, recovered metals or metal salts have the best potential for reassimilation into a manufactured product. The metal finishing and electronic industries for the most part generate metal wastes in a condition not so readily assimilated into product manufacturing, so that off-site recovery may be the preferred option.

High disposal costs combined with rigorous enforcement of pollution emission regulations provide the principal incentives for most generators, large or small. The treatment companies which collect and process wastes may also have an incentive to recycle to minimize the residues requiring off-site disposal. Off-site disposal of hazardous wastes is a big business and the major treatment companies operating nationwide, such as Chemical Waste Management, Browning Ferris, Inc./CECOS, Zimpro, Rollins Environmental Services, SCA Chemical Services, Ensco Inc., Chem Clear Inc., and ERT, have annual sales ranging from $2 million to $150 million. Large companies are most likely to proceed with on-site disposal solutions. Small hazardous waste generators are most dependent on off-site disposal.

In addition to the large companies that undertake to treat all organic and inorganic wastes, there are a number of companies that provide hardware, and design systems and offer consulting services in a variety of separation processes relevant to the treatment of industrial wastes (not all of these address metal recovery specifically). Prominent separation processes are electrowinning, ion exchange, membrane and magnetic separations, and chemical recovery. Some of these companies are listed in Table 10.1. Listings of firms providing these services can be found in an appropriate directory.[17]

Successful disposal of metal finishing industry wastes for small generators by incorporation of metal resource recovery may require a combination of private, commercial, and state action for creation of an infrastructure that currently does not exist. One route for creation of this infrastructure designed particularly for the off-site disposal problems of both small and large generators is through development of centralized regional waste treatment (CWT) facilities. Such facilities are used commonly in Japan and in some European countries, such as Germany. The need for CWT facilities in the United States is growing rapidly, but such facilities are currently only undergoing evaluation. A CWT facility must be designed to handle the volume of metal wastes necessary to make recovery efficient and to provide economies of scale. Several analyses of the economic viability of CWT facilities have been conducted and found positive.[8 to 13, 21, 22] An important consideration in the design of CWT facilities is provision for secure disposal, by some type of land disposal, of the innocuous, irreducible residues left from the destruction or conversion of the hazardous materials and recovery of resources. There are several promising processes for production of inert materials for land disposal to the extent that is acceptable under state and federal regulation. Examples that can be cited are the following processes: Solinoc, Solid Tek, Stablex, ChemFix, and Poz-o-Tec.[30]

10.4 FUTURE DEVELOPMENTS

The incentives for metal recovery from metal wastes lie principally in obtaining reductions in waste disposal costs, improving compliance with pollution standards, and conservation of metals of economic significance and strategic value.

Table 10.1 Companies Providing Services Relevant to Metal Recovery from Industrial Wastes

Company	Location	Adsorption	Electro-winning	Ion exchange	Membrane separation	Magnetic separation
Alcoa Separations Tech. Div.	Warrendale, PA	Alumina adsorbant	X	X	X	
CX Technologies Inc.	Yorkville, IL	Cellulose xanthate adsorbant				
Westvaco	Covington, VA	Carbon adsorbant				
Chemlec	Whitby, Ontario		X			
Eltech	Chardon, OH		X			
Ecotec Ltd.	Pickering, Ontario		X	X		
Baker Brothers/Systems	Stoughton, MA		X			
Covo Finish Co., Inc.	Burrillville, RI		X			
Metal Recovery Tech., Inc.	Cumberland, RI		X			
ANDCO	Amherst, NY		X			
MEMTEX	Billerica, MA			X	X	
Illinois Water Treatment Co.	Rockford, IL			X		
Douglas Water and Metal Recycling System	Eden Prairie, MN			X		
RAI Research Corporation	Hauppage, NY			X	X	

Table 10.1 continued

Company	Location	Adsorption	Electro-winning	Ion exchange	Membrane separation	Magnetic separation
Met-Pro Corp., Sethco Div.	Hauppage, NY			X		
Amicon	Danvers, MA				X	
duPont	Wilmington, DE				X	
Ionics	Watertown, MA				X	
Osmonics	Minnetonka, MN				X	
Parkans Int., Inc.	Houston, TX				X	
Polymetrics	Boston, MA				X	
HPD Inc.	Naperville, IL				X	
Bend Research, Inc.	Bend, OR				X	
Barnstead Co.	Newton, MA				X	
Manchester Corp.	Harvard, MA				X	
IGC (Intermagnetics Gen.)	Guilderland, NY					X

Total metal waste volumes, even on a regional basis, do not amount to a very significant fraction of the current production volumes and use volumes and are not easily perceived as new metal sources. Furthermore, the generation sources are numerous, highly diversified, and geographically dispersed. The most obvious economic advantages accrue to the generator-operated metal recovery facility. The generator has the most to gain by having control over process conditions that permit recovery of "metal by-products" of marketable value from effluents with the simplest composition. Treatment companies with the capability to incorporate resource recovery into existing waste metal disposal processes are second only to the generators in having the opportunity to benefit from the economic credits from recovered metals. In addition to establishing efficient separation/concentration technologies, treatment companies can play an important role in developing a collection/transportation infrastructure for concentration of the metals for transfer to potential markets.

A positive climate favoring recycling is only feasible if several conditions are met:

- positive corporate policy
- government regulations with positive incentives
- general acceptance and market for recycled materials
- active demand for strategic metals
- significant volume for recovered materials
- good quality and consistency in production rate
- close proximity of source to potential market
- economical, technically feasible recovery technology
- disengagement of hazardous waste classification from recovered materials
- active marketing of recovered materials

Successful recycling is only likely to occur in any given case if most of these factors are present.

References

1. Wadsworth, M. E. and F. T. Davis. *Unit Processing Hydrometallurgy* (New York: Gordon & Breach, Science Publishers, Inc., 1964).
2. Burkin, A. R. *Chemistry of Hydrometallurgical Processes* (Princeton, NJ: D. Van Nostrand Company, 1966).
3. Pehlke, R. D. *Unit Processes of Extractive Metallurgy* (New York: American Elsevier, 1973).
4. Bautista, R. G. "Hydrometallurgy." in *Advances in Chemical Engineering*, Vol. 9, T. B. Drew, G. R. Cokelet, J. W. Hoopes, Jr., and T. Vermeulen, Eds. (New York: Academic Press, 1974).
5. Gupta, C. K., and T. K. Mukherjee, Eds. *Hydrometallurgy in Extraction Processes*, Vols. 1 and 2 (Boca Raton, FL: CRC Press, Inc., 1990).
6. Habashi, F. "Hydrometallurgy," in *Principles of Extractive Metallurgy*, Vol. 2 (New York: Gordon & Breach, Science Publishers, 1970).
7. Rouseau, R. W. *Handbook of Separation Process Technology*. (New York: Wiley-Interscience Publishers, 1987).
8. Twidwell, L. G. "Metal Recovery from Metal Hydroxide Sludges." EPA 600/52-85/128 NTIS PB 86-157294 (1986).
9. Ball, R. O., P. L. Buckingham, and S. Mahfoud. "Economic Feasibility of a State-Wide Hydrometallurgical Recovery Facility," in *Metals Speciation and Recovery*, J. W. Patterson and R. Passino, Eds. (Chelsea, MI: Lewis Publishers, Inc., 1987), p. 684.
10. Stinson, M. K. "Group Treatment of Multicompany Plating Waste, Tontin, MA Silver Project." EPA-600/2-710-102; PB 80-102825 (1979).
11. "Environmental Pollution Control Alternatives: Centralized Waste Treatment Alternatives for the Electroplating Industry." EPA 625/5-81-017 (1981).
12. Pope-Reid Associates, Inc. "Feasibility Study—Central Recovery Facility for Minnesota Department of Energy, Planning & Development." (1982).

13. Gupta, A., et al. *A Central Metal Recovery Facility*, Dept. of Chem Engr. (Princeton, NJ: Princeton University Press, October, 1983).
14. Herndon, R.C., Ed. *Proceedings of the First National Conference on Waste Exchange*. (Tallahassee, FL: The Florida State University, Florida Chamber of Commerce and U.S. EPA, March, 1983).
15. Hill, R. P., EPA Office of Solid Waste. United States Waste Exchanges (1982).
16. Herndon, R. C. and E. D. Purdam, Eds. "Proceedings of the 2nd National Conference on Waste Exchange," EPA/ Florida State University, March 1985.
17. *Chemical Week Directory of Worldwide CPI Services* (issued annually).
18. Coleman, R. T. et al. "Sources and Treatment of Wastewater in the Nonferrous Metals Industry." EPA-600/ 2–80–074; NTIS PB80–196118 (1980).
19. Clark-McGlennon Associates, Inc. and Waste Resource Associates. "Managing Metal Hydroxide Sludge in Connecticut." (Report for U.S. EPA Region I Waste Management Division and Connecticut Department of Environmental Protection [1982]).
20. American Electroplaters Society and U.S. EPA. Third Conference on Advanced Pollution Control for the Metal Finishing Industry, April, 1980. EPA 600/2–81–028 (1981).
21. EPA Fourth Conference on Advanced Pollution Control for the Metal Finishing Industry. EPA 600/10–82–022; NTIS PB 83–165803 (1982).
22. Environmental Regulations and Technology – The Electroplating Industry. EPA 625/10–80–001 (1980).
23. Maugh T. H., II. "Burial is Last Resort for Hazardous Wastes," *Science* 204:1295 (1979).
24. Graff, M. "Selling Waste Can Check Rising Disposal Costs," *Chem. Engr.* (October 4, 1982), pp. 310–43.
25. Conway, R. A. and R. D. Ross. *Handbook of Industrial Waste Disposal* (New York: Van Nostrand Reinhold Company, 1980).

26. Economic Analysis of Effluent Guidelines, The Metal Finishing Industry. EPA 230/1–74–032 (1974).
27. Wapora, Inc. "Assessment of Industrial Hazardous Waste Practice, Special Machinery Manufacturing Industries." EPA SW-141C; PB 265981 (1977).
28. "Environmental Pollution Control Alternatives: Economics of Wastewater Treatment Alternatives for the Electroplating Industry." EPA 625/5–710–016 (1979).
29. "Control and Treatment Technology for the Metal Finishing Industry: In-Plant Changes." EPA 625/8–82–008 (1982).
30. Pojasek, R. B. "Solid-Waste Disposal: Solidification," *Chem. Eng.* (August 13, 1979), pp. 141–146.

Appendix A

EPA and State Hazardous Waste Contacts for Assistance

RCRA/Superfund Hotline 1-800-424-9346 (In Washington, D.C.:382-3000)	EPA Small Business Ombudsman Hotline 1-800-368-5888 (In Washington, D.C.:557-1938)	National Response Center 1-800-424-8802 (In Washington, D.C.:426-2675)

Regions	Regions	Regions	Regions
4—Alabama	5—Indiana	9—Nevada	4—Tennessee
10—Alaska	7—Iowa	1—New Hampshire	6—Texas
9—Arizona	7—Kansas	2—New Jersey	8—Utah
6—Arkansas	4—Kentucky	6—New Mexico	1—Vermont
9—California	6—Louisiana	2—New York	3—Virginia
8—Colorado	1—Maine	4—North Carolina	10—Washington
1—Connecticut	3—Maryland	8—North Dakota	3—West Virginia
3—Delaware	1—Massachusetts	5—Ohio	5—Wisconsin
3—D.C.	5—Michigan	6—Oklahoma	8—Wyoming
4—Florida	5—Minnesota	10—Oregon	9—American Samoa
4—Georgia	4—Mississippi	3—Pennsylvania	9—Guam
9—Hawaii	7—Missouri	1—Rhode Island	2—Puerto Rico
10—Idaho	8—Montana	4—South Carolina	2—Virgin Islands
5—Illinois	7-Nebraska	8—South Dakota	

U.S. EPA REGIONAL OFFICES

EPA Region I
State Waste Programs Branch
JFK Federal Building
Boston, MA 02203
(617) 223-3468
Connecticut, Massachusetts, Maine,
  New Hampshire, Rhode Island, Vermont

EPA Region II
Air and Waste Management Division
26 Federal Plaza
New York, NY 10278
(212) 264-5175
New Jersey, New York, Puerto Rico,
  Virgin Islands

EPA Region III
Waste Management Branch
841 Chestnut Street
Philadelphia, PA 19107
(215) 597-9336
Delaware, Maryland, Pennsylvania,
  Virginia, West Virginia,
  District of Columbia

EPA Region IV
Hazardous Waste Management Division
345 Courtland Street, N.E.
Atlanta, GA 30365
(404) 347-3016
Alabama, Florida, Georgia, Kentucky, Mississippi,
  North Carolina, South Carolina, Tennessee

EPA Region V
RCRA Activities
230 South Dearborn Street
Chicago, IL 60604
(312) 353-2000
Illinois, Indiana, Michigan,
    Minnesota, Ohio, Wisconsin

EPA Region VI
Air and Hazardous Materials Division
1201 Elm Street
Dallas, TX 75270
(214) 767-2600
Arkansas, Louisiana, New Mexico,
    Oklahoma, Texas

EPA Region VII
RCRA Branch
726 Minnesota Avenue
Kansas City, KS 66101
(913) 236-2800
Iowa, Kansas, Missouri, Nebraska

EPA Region VIII
Waste Management Division (8HWM-ON)
One Denver Place
999 18th Street, Suite 1300
Denver, CO 80202-2413
(303) 293-1502
Colorado, Montana, North Dakota,
    South Dakota, Utah, Wyoming

EPA Region IX
Toxics and Waste Management Division
215 Fremont Street
San Francisco, CA 94105
(415) 974-7472
Arizona, California, Hawaii,
  Nevada, American Samoa, Guam,
  Trust Territories of the Pacific

EPA Region X
Waste Management Branch—MS-530
1200 Sixth Avenue
Seattle, Washington 98101
(206) 422-2777
Alaska, Idaho, Oregon, Washington

STATE HAZARDOUS WASTE MANAGEMENT AGENCIES

Alabama
Alabama Department of
Environment Management
Land Division
1751 Federal Drive
Montgomery, AL 36130
(205) 271-7730

Alaska
Department of Environmental Conservation
P.O. Box 0
Juneau, AL 99811

Program Manager: (907) 465-2666
Northern Regional Office
(Fairbanks): (907) 452-1714
South-Central Regional Office
(Anchorage): (907) 274-2533
Southeast Regional Office
(Juneau): (907) 789-3151

American Samoa
Environmental Quality Commission
Government of American Samoa
Pago Pago, American Samoa 96799
Overseas Operator
(Commercial Call (684) 663-4116)

Arizona
Arizona Department of Health Services
Office of Waste and Water Quality
2005 North Central Avenue
  Room 304
Phoenix, AZ 85004
Hazardous Waste Management:
(602) 255-2211

Arkansas
Department of Pollution Control and Ecology
Hazardous Waste Division
P.O. Box 9583
8001 National Drive
Little Rock, AR 72219
(501) 562-7444

California
Department of Health Services
Toxic Substances Control Division
714 P Street, Room 1253
Sacramento, CA 95814
(916) 324-1826

State Water Resources Control Board
Division of Water Quality
P.O. Box 100
Sacramento, CA 95801
(916) 322-2867

Colorado
Colorado Department of Health
Waste Management Division
4210 E. 11th Avenue
Denver, CO 80220
(303) 320-8333 Ext. 4364

Connecticut
Department of Environmental Protection
Hazardous Waste Management
Section
State Office Building
165 Capitol Avenue
Hartford, CT 06106
(203) 566-8843, 8844

Connecticut Resource Recovery Authority
179 Allyn Street, Suite 603
Professional Building
Hartford, CT 06103
(203) 549-6390

Delaware
Department of Natural Resources
and Environmental Control
Waste Management Section
P.O. Box 1401
Dover, DE 19903
(302) 736-4781

District of Columbia
Department of Consumer and Regulatory Affairs
Pesticides and Hazardous Waste
Materials Division
Room 114
5010 Overlook Avenue, S.W.
Washington, DC 20032
(202) 767-8414

Florida
Department of Environmental Regulation
Solid and Hazardous Waste Section
Twin Towers Office Building
2600 Blair Stone Road
Tallahassee, FL 32301
RE: SQG's
(904) 488-0300

Georgia
Georgia Environmental Protection Division
Hazardous Waste Management
Program
Land Protection Branch
Floyd Towers East, Suite 1154
205 Butler Street, S.E.
Atlanta, GA 30334
(404) 656-2833
Toll Free: (800) 334-2373

Guam
Guam Environmental Protection Agency
P.O. Box 2999
Agana, GU 96910
Overseas Operator
(Commercial Call (671) 646-7579)

Hawaii
Department of Health
Environmental Health Division
P.O. Box 3378
Honolulu, HI 96801
(808) 548-4383

Idaho
Department of Health and Welfare
Bureau of Hazardous Materials
450 West State Street
Boise, ID 83720
(208) 334-5879

Illinois
Environmental Protection Agency
Division of Land Pollution Control
2200 Churchill Road, #24
Springfield, IL 62706
(217) 782-6761

Indiana
Department of Environmental Management
Office of Solid and Hazardous Waste
105 South Meridian
Indianapolis, IN
(317) 232-4535

Iowa
U.S. EPA Region VII
Hazardous Materials Branch
726 Minnesota Avenue
Kansas City, KS 66101
(913) 236-28888
Iowa RCRA Toll Free:
(800) 223-0425

Kansas
Department of Health and Environment
Bureau of Waste Management
Forbes Field, Building 321
Topeka, KS 66620
(913) 862-9360 Ext. 292

Kentucky
Natural Resources and
Environmental Protection Cabinet
Division of Waste Management
18 Reilly Road
Frankfort, KY 40601
(502) 564-6716

Louisiana
Department of Environmental Quality
Hazardous Waste Division
P.O. Box 44307
Baton Rouge, LA 70804
(504) 342-1227

Maine
Department of Environmental Protection
Bureau of Oil and Hazardous
Materials Control
State House Station #17
Augusta, ME 04333
(207) 289-2651

Maryland
Department of Health and Mental Hygiene
Maryland Waste Management
Administration
Office of Environmental Programs
201 West Preston Street, Room A3
Baltimore, MD 20201
(301) 225-5709

Massachusetts
Department of Environmental Quality Engineering
Division of Solid and Hazardous Waste
One Winter Street, 5th Floor
Boston, MA 02108
(617) 292-5589
(617) 292-5851

Michigan
Michigan Department of Natural Resources
Hazardous Waste Division
Waste Evaluation Unit
Box 30028
Lansing, MI 48909
(517) 373-2730

Minnesota
Pollution Control Agency
Solid and Hazardous Waste Division
1935 West County Road, B-2
Roseville, MN 55113
(612) 296-7282

Mississippi
Department of Natural Resources
Division of Solid and Hazardous
Waste Management
P.O. Box 10385
Jackson, MS 39209
(601) 961-5062

Missouri
Department of Natural Resources
Waste Management Program
P.O. Box 176
Jefferson City, MO 65102
(314) 751-3176
Missouri Hotline:
(800) 334-6946

Montana
Department of Health and
Environmental Sciences
Solid and Hazardous Waste Bureau
Cogswell Building, Room B-201
Helena, MT 59620
(406) 444-2821

Nebraska
Division of Environmental Control
Waste Management Hazardous Section
P.O. Box 94877
Lincoln, NE 68509
(402) 471-2186

Nevada
Division of Environmental Protection
Waste Management Program
Capitol Complex
Carson City, NV 89710
(702) 885-4670

New Hampshire
Department of Health and Human Services
Division of Public Health Services
Office of Waste Management
Health and Welfare Building
Hazen Drive
Concord, NH 03301-6527
(603) 271-4608

New Jersey
Department of Environmental Protection
Division of Waste Management
32 East Hanover Street, CN-028
Trenton, New Jersey 08625
Hazardous Waste Advisement
Program: (609) 292-8341

New Mexico
Environmental Improvement Division
Ground Water and Hazardous
Waste Bureau
Hazardous Waste Section
P.O. Box 968
Santa Fe, NM 87504-0968
(505) 827-2922

New York
Department of Environmental Conservation
Bureau of Hazardous Waste Operations
50 Wolf Road, Room 209
Albany, NY 12233
(518) 457-0530
SQG Hotline: (800) 631-0666

North Carolina
Department of Human Resources
Solid and Hazardous Waste
Management Branch
P.O. Box 2091
Raleigh, NC 27602
(919) 733-2178

North Dakota
Department of Health
Division of Hazardous Waste
Management and Special Studies
1200 Missouri Avenue
Bismarck, ND 58502-5520
(701) 224-2366

**Northern Mariana Islands,
Commonwealth of**
Department of Environmental and
Health Services
Division of Environmental Quality
P.O. Box 1304
Saipan, Commonwealth of
Mariana Islands 96950
Overseas call (670) 234-6984

Ohio
Ohio EPA
Division of Solid and Hazardous
Waste Management
361 East Broad Street
Columbus, OH 43266-0558
(614) 466-7220

Oklahoma
Waste Management Service
Oklahoma State Department of
Health
P.O. Box 53551
Oklahoma City, OK 73152
(405) 271-5338

Oregon
Hazardous and Solid Waste Division
P.O. Box 1760
Portland, OR 97207
(503) 229-6534
Toll Free: (800) 452-4011

Pennsylvania
Bureau of Waste Management
Division of Compliance Monitoring
P.O. Box 2063
Harrisburg, PA 17120
(717) 787-6239

Puerto Rico
Environmental Quality Board
P.O. Box 11488
Santurce, PR 00910-1488
(809) 723-8184

-or-

EPA Region II
Air and Waste Management Division
26 Federal Plaza
New York, NY 10278
(212) 264-5175

Rhode Island
Department of Environmental Management
Division of Air and Hazardous
Materials
Room 204, Cannon Building
75 Davis Street
Providence, RI 02908
(401) 277-2797

South Carolina
Department of Health and
Environmental Control
Bureau of Solid and Hazardous
Waste Management
2600 Bull Street
Columbia, SC 29201
(803) 734-5200

South Dakota
Department of Water and Natural Resources
Office of Air Quality and Solid Waste
Foss Building, Room 217
Pierre, SD 57501
(605) 773-3153

Tennessee
Division of Solid Waste Management
Tennessee Department of Public
Health
701 Broadway

Nashville, TN 37219-5403
(615) 741-3424

Texas
Texas Water Commission
Hazardous and Solid Waste Division
Attn: Program Support Section
1700 North Congress
Austin, TX 78711
(512) 463-7761

Utah
Department of Health
Bureau of Solid and Hazardous
Waste Management
P.O. Box 16700
Salt Lake City, UT 84116-0700
(801) 538-6170

Vermont
Agency of Environmental Conservation
103 South Main Street
Waterbury, VT 05676
(802) 244-8702

Virgin Islands
Department of Conservation and
Cultural Affairs
P.O. Box 4399
Charlotte Amalie, St. Thomas
Virgin Islands 00801
(809) 774-3320

-or-

EPA Region II
Air and Waste Management Division
26 Federal Plaza

New York, NY 10278
(212) 264-5175

Virginia
Department of Health
Division of Solid and Hazardous
Waste Management
Monroe Building, 11th Floor
101 North 14th Street
Richmond, VA 23219
(804) 225-2667
Hazardous Waste Hotline:
(800) 552-2075

Washington
Department of Ecology
Solid and Hazardous Waste Program
Mail Stop PV-11
Olympia, WA 98504-8711
(206) 459-6322
In-State: 1-800-633-7585

West Virginia
Division of Water Resources
Solid and Hazardous Waste/
Ground Water Branch
1201 Greenbrier Street
Charleston, WV 25311

Wisconsin
Department of Natural Resources
Bureau of Solid Waste Management
P.O. Box 7921
Madison, WI 53707
(608) 266-1327

Wyoming
Department of Environmental Quality
Solid Waste Management Program
122 West 25th Street
Cheyenne, WY 82002
(307) 777-7752

-or-

EPA Region VIII
Waste Management Division
(8HWM-ON)
One Denver Place
999 18th Street
Suite 1300
Denver, CO 80202-2413
(303) 293-1502

Appendix B

Sources of Data for Chapter 9

A thorough attempt has been made to identify sources of data to support the analysis presented in Chapter 9. Listed below are various data sources which contributed significantly to gaining an understanding of the economics of the metal recovery industry. When appropriate, the reliability and various limitations of the data sources are discussed.

A. Waste Stream Volume

- *1989 State Capacity Assurance Plans (CAPs)*
  - CAPs are available to the public at the RCRA docket room at 401 M Street, SW, Washington, DC. The phone number is (202)475-9327.
  - Each state was required by Congress to submit to the EPA a CAP by 17 October 1989.
  - CAPs include actual data for 1987 and projected data for the years 1989, 1995 and 2009.
  - They include data on the volumes of hazardous waste generated in, imported to, exported from, and managed in each state by waste type.
  - Data are sometimes unreliable and inconsistent between states, complicating the comparison or consolidation of reported volumes across states.
- *1985 National Report of Hazardous Waste Generators and Treatment, Storage and Disposal Facilities Regulated Under RCRA*
  U.S. Environmental Protection Agency, Office of Solid Waste, Office of Policy, Planning, and Information, Wash-

ington DC, March 1989 (Document number: EPA/530-SW-89–033B).
- −Published biennially.
- −Contains data on hazardous waste-generating companies and volumes of waste generated, as well as data on volumes of waste managed (equivalent to waste generated plus waste imported less waste exported) by various methods.
- −Most data aggregated across all states.
- −Provides some individual state data on waste generation and TSDFs.

B. Metal Prices

- London Metal Exchange prices published daily in *The Financial Times*.
- Prices from the London Metal Exchange and other sources are published in *Metals Week*.

C. Metal Recycling Volumes

- *Mineral Commodity Summaries 1990*
  U.S. Bureau of Mines, Washington, DC, January 1991
  - −Contains production and consumption data on various individual metals.
  - −Contains some data on recycled quantities of metals.
  - −Contains little summary data.
- *Facts 1989 Yearbook*
  - −The Institute of Scrap Recycling Industries, Inc. (ISRI), Washington, DC, 1990.
  - −Contains consumption, export, import, recycled market share, and price data for various recycled metals.
  - −Does not utilize consistent recycling formulas for generation of recycled market share, making comparisons between certain metals difficult.

D. Regulatory Issues

- *The Resource Conservation and Recovery Act of 1976 (RCRA)*
  90 Stat. 95; 42 U.S.C. 6901 et seq.
- *The Comprehensive Environmental Response, Compensation and Liability Act of 1980 (CERCLA)*
  94 Stat. 2767; 42 U.S.C. 9601 et seq.
- *The Nation's Hazardous Waste Management Program at a Crossroads; The RCRA Implementation Study*
  U.S. Environmental Protection Agency, Office of Solid Waste and Emergency Response, July 1990
- *Land Disposal Inventory for First Third Scheduled Wastes, Federal Register*, V.53(159), August 17, 1988
- *Land Disposal Restrictions for Second Third Scheduled Wastes, Federal Register*, V.54(120), June 23, 1989
- *Land Disposal Requirements for Third Third Scheduled Wastes, Federal Register*, V.54(224), November 22, 1989

E. Cost Information

- *Economic Feasibility of a State-Wide Hydrometallurgical Recovery Facility*, Ray O. Ball, et al., *Metals Speciation, Separation and Recovery*, Lewis Publishers, 1987
  - Contains information concerning capital costs associated with a metal recovery venture.
- *1986–1987 Survey of Selected Firms in the Commercial Hazardous Waste Management Industry*
  U.S. Environmental Protection Agency, Office of Policy Analysis
  - Contains information concerning transportation costs for hazardous wastes.

F. Contingent Liability

- *Extent of the Hazardous Release Problem and Future Funding Needs: CERCLA Section 301(a)(1)(C) Study*
  U.S. Environmental Protection Agency, Office of Solid Waste and Emergency Response, December 1984

- —Contains data on the average cleanup cost of a Superfund site.
- 40 CFR Part 300, National Priorities List for Uncontrolled Hazardous Waste Sites
  - —Contains data on the average cleanup cost of a Superfund site.
- ROD Annual Report: FY 1989
  U.S. Environmental Protection Agency, Office of Emergency and Remedial Response, April 1990
  - —Contains data on volume and cleanup costs for Superfund sites.

Appendix C

Waste Exchanges Operating in North America (May 1988)

Alabama Waste Exchange
Mr. William J. Herz
The University of Alabama
P.O. Box 870203
Tuscaloosa, AL 35487-0203
(205) 349-5889; FAX (205) 384-8573

Great Lakes Waste Exchange
Ms. Kay Ostrowski
400 Ann St., N.E. Suite 201-A
Grand Rapids, MI 49504-2054
(616) 363-3262

California Waste Exchange
Mr. Robert McCormick
Dept. of Health Services
Toxic Substances Control Div.
Alternative Technology Section
P.O. Box 942732
Sacramento, CA 94234-7320
(916) 324-1807

Manitoba Waste Exchange
Mr. James Ferguson
c/o Biomass Energy Institute, Inc.
1329 Niakwa Rd.
Winnipeg, Manitoba
Canada R2J 3T4
(204) 257-3891

New Hampshire Waste Exchanges
Mr. Gary J. Olson c/o NHRRA
P.O. Box 721
Concord, NY 03301
(603) 224-6996

Southern Waste Information Exchange
Mr. Eugene B. Jones
P.O. Box 960
Tallahassee, FL 32202
(800) 441-SWIX; (904) 644-5516
FAX (904) 574-6704

Pacific Materials Exchange
Mr. Bob Smee
South 3707 Godfrey Blvd.
Spokane, WA 99204
(509) 623-4244

Enstar Corporation*
Mr. J. T. Engser
P.O. Box 189
Latham, NY 12110
(518) 785-0470

British Columbia Waste Exchange
Mr. Lynn Deegan
2150 Maple Street
Vancouver, BC
Canada, V6J 3T3
(604) 731-7222

Industrial Materials Exchange
(IMEX)
Mr. Jerry Henderson
172 20th Avenue
Seattle, WA 98122
(206) 296-4633; FAX (206) 296-0188

Canadian Waste Materials Exchange
ORTECH International
Dr. Robert Laughlin
2395 Speakman Drive
Mississauga, Ontario
Canada L5K 1B3
(416) 822-4111 ext 265

Montana Industrial Waste Exchange
Mr. Don Ingles
Montana Chamber of Commerce
P.O. Box 1730
Helena, MT 59624
(406) 442-2405

Southeast Waste Exchange
Ms. Mary McDaniel
Urban Institute
UNCC Station
Charlotte, NC 28223
(704) 547-2307

Ontario Waste Exchange
ORTECH International
Ms. Linda Varangu
2395 Speakman Drive
Mississauga, Ontario
Canada L5K 1B3
(416) 822-4111 ext 512

Peel Regional Waste Exchange
Mr. Glen Milbury
Regional Municipality of Peel
10 Peel Center Drive
Brampton, Ontario
Canada L6T 4B9
(416) 791-9400

Alberta Waste Materials Exchange
Mr. William C. Kay
Alberta Research Council
P.O. Box 8330
Postal Station F
Edmonton, Alberta
Canada T6H 5X2
(403) 450-5408

Indiana Waste Exchange
Ms. Susan Scrogham
P.O. Box 1220
Indianapolis, IN 46206
(317) 634-2142

Canadian Chemical Exchange*
Mr. Philippe LaRoche
PO Box 1135
Ste. Adele, Quebec
Canada J0R 1L0
(514) 229-6511

Industrial Waste Information Exchange
Mr. William E. Payne
New Jersey Chamber of Commerce
5 Commerce Street
Newark, NJ 07102
(201) 623-7070

RENEW
Ms. Cheryl Wilson
Texas Water Commission
P.O. Box 13078
Austin, TX 78711-3087
(512) 463-7773; FAX (512) 463-8317

Northeast Industrial Waste Exchange
Mr. Lewis Cutler
90 Presidential Plaza, Suite 122
Syracuse, NY 13202
(315) 422-6572; FAX (315) 422-9051

Wastelink, Division of Tencon Inc
Ms. Mary E. Malotke
140 Wooster Pike
Milford, OH 45150
(513) 248-0012; FAX (513) 248-1094

*For Profit Waste Information Exchange.

Index to References

This index is applicable to chapters 5, 7, and 8 and cross-references are specific to the metal and to the principal separation process described. The references for all the other chapters are listed at the end of the given section or chapter and are not indexed.

Index by Process

key: aa-bb-nnn
aa = Metal; bb = Process Reference; nn = Reference Number

Adsorption: Ag–A–3, Ag–A–5, Ag–A–7, Ag–A–8, Ag–A–10, Ag–A–17, Ag–A–22, Al–A–12, Al–A–22, As–A–17, Be–A–17, Cd–A–3, Cd–A–5, Cd–A–17, Cd–A–22, Co–A–13, Co–A–20, Cr–A–17, Cu–A–2, Cu–A–3, Cu–A–5, Cu–A–8, Cu–A–10, Cu–A–11, Cu–A–12, Cu–A–13, Cu–A–14, Cu–A–15, Cu–A–16, Cu–A–17, Cu–A–20, Cu–A–22, Fe–A–12, Hg–A–8, Hg–A–17, Hg–A–22, Mn–A–13, Ni–A–1, Ni–A–3, Ni–A–4, Ni–A–5, Ni–A–6, Ni–A–9, Ni–A–13, Ni–A–16, Ni–A–17, Ni–A–19, Pb–A–17, Pd–A–8, Pd–A–14, Pd–A–19, Sb–A–17, Sn–A–17, Ti–A–17, Ti–A–21, U–A–21, Zn–A–4, Zn–A–12, Zn–A–14, Zn–A–17

Biological: Ag–B–15, Ag–B–16, Al–B–15, ^{241}Am–B–12, As–B–2, Cd–B–5, Cd–B–7, Cd–B–14, Cd–B–15, Cd–B–16, Ce–B–12, Co–B–12, Co–B–15, Cr–B–15, Cs–B–2, Cs–B–12, ^{134}Cs–B–12, Cu–B–4, Cu–B–5, Cu–B–7, Cu–B–8, Cu–B–9, Cu–B–10, Cu–B–11, Cu–B–12, Cu–B–13,

253

Cu–B–14, Cu–B–15, Cu–B–16, Fe–B–2, Hg–B–2, Mn–B–2, Mo–B–2, Mo–B–3, Ni–B–2, Ni–B–7, Ni–B–14, Ni–B–15, Pb–B–2, Pb–B–5, Pb–B–14, Pb–B–15, Pb–B–16, ^{242}Pu–B–12, Ra–B–2, Ra–B–3, Se–B–2, Se–B–3, Sn–B–2, Sr–B–12, ^{85}Sr–B–12, U–B–2, U–B–13, V–B–3, Zn–B–7, Zn–B–14, Zn–B–15, Zn–B–16, Zr–B–12

Cementation: Ag–C–1, Ag–C–14, Ag–C–16, Al–C–1, Al–C–5, Al–C–8, Al–C–20, Al–C–21, Al–C–22, As–C–12, Au–C–1, Cd–C–1, Cd–C–4, Cr–C–11, Cu–C–1, Cu–C–2, Cu–C–3, Cu–C–5, Cu–C–6, Cu–C–7, Cu–C–8, Cu–C–9, Cu–C–10, Cu–C–11, Cu–C–12, Cu–C–13, Cu–C–14, Cu–C–15, Cu–C–17, Cu–C–18, Cu–C–19, Cu–C–20, Cu–C–21, Fe–C–1, Fe–C–2, Fe–C–3, Fe–C–6, Fe–C–7, Fe–C–8, Fe–C–9, Fe–C–10, Fe–C–11, Fe–C–17, Fe–C–18, Fe–C–19, Fe–C–22, Ga–C–1, Mg–C–4, Pb–C–1, Pb–C–15, Pd–C–14, Pd–C–16, Sb–C–12, Sn–C–12, Zn–C–1, Zn–C–11, Zn–C–12, Zn–C–13, Zn–C–15, Zn–C–16

Electrowinning: Ag–E–38, Ag–E–39, Ag–E–42, Ag–E–77, Ag–E–85, Ag–E–107, Ag–E–125, Au–E–38, Au–E–39, Au–E–52, Au–E–107, Au–E–121, Au–E–125, Cd–E–34, Cd–E–38, Cd–E–39, Cd–E–51, Cd–E–55, Cd–E–60, Cd–E–61, Cd–E–67, Cd–E–71, Cd–E–72, Cd–E–113, Cu–E–29, Cu–E–31, Cu–E–32, Cu–E–34, Cu–E–35, Cu–E–37, Cu–E–38, Cu–E–39, Cu–E–40, Cu–E–43, Cu–E–44, Cu–E–46, Cu–E–47, Cu–E–48, Cu–E–49, Cu–E–51, Cu–E–52, Cu–E–53, Cu–E–55, Cu–E–56, Cu–E–57, Cu–E–58, Cu–E–60, Cu–E–62, Cu–E–63, Cu–E–64, Cu–E–65, Cu–E–66, Cu–E–67, Cu–E–69, Cu–E–70, Cu–E–72, Cu–E–73, Cu–E–74, Cu–E–76, Cu–E–77, Cu–E–79, Cu–E–81, Cu–E–83, Cu–E–84, Cu–E–87, Cu–E–88, Cu–E–91, Cu–E–92, Cu–E–93, Cu–E–94, Cu–E–95, Cu–E–96, Cu–E–99, Cu–E–103, Cu–E–104, Cu–E–105, Cu–E–106, Cu–E–107, Cu–E–108, Cu–E–111, Cu–E–116, Cu–E–117, Cu–E–118, Cu–E–120, Cu–E–122, Cu–E–124, Cu–E–125, Cu–E–126, Cu–E–127, Cu–E–128, Cu–E–131, Cu–E–132, Cu–E–134, Nb–E–112,

Zn-F-31, Zn-F-33, Zn-F-36, Zn-F-39, Zn-F-40, Zn-F-51, Zn-F-59, Zn-F-60, Zn-F-63, Zn-F-75

Ion Exchange: Ag-I-21, Ag-I-24, Ag-I-35, Ag-I-45, Ag-I-48, Ag-I-56, Ag-I-75, Al-I-10, Au-I-5, Au-I-35, Au-I-48, Au-I-54, Au-I-56, Au-I-94, Bi-I-5, Cd-I-21, Cd-I-42, Cd-I-45, Cd-I-72, Cd-I-97, Ce-I-87, Co-I-5, Co-I-25, Co-I-43, Co-I-51, Co-I-68, Co-I-69, Co-I-78, Co-I-80, Co-I-96, Co-I-98, Cr-I-18, Cr-I-23, Cr-I-25, Cr-I-56, Cr-I-71, Cr-I-97, Cu-I-19, Cu-I-21, Cu-I-22, Cu-I-24, Cu-I-25, Cu-I-27, Cu-I-29, Cu-I-32, Cu-I-35, Cu-I-38, Cu-I-40, Cu-I-41, Cu-I-42, Cu-I-43, Cu-I-44, Cu-I-45, Cu-I-46, Cu-I-47, Cu-I-48, Cu-I-49, Cu-I-51, Cu-I-53, Cu-I-54, Cu-I-59, Cu-I-60, Cu-I-61, Cu-I-62, Cu-I-63, Cu-I-66, Cu-I-67, Cu-I-68, Cu-I-69, Cu-I-71, Cu-I-72, Cu-I-74, Cu-I-76, Cu-I-78, Cu-I-81, Cu-I-82, Cu-I-86, Cu-I-88, Cu-I-89, Cu-I-91, Cu-I-92, Cu-I-97, Cu-I-98, Cu-I-99, Cu-I-100, Cu-I-101, Fe-I-5, Fe-I-18, Fe-I-51, Fe-I-56, Fe-I-68, Fe-I-80, Fe-I-96, Fe-I-97, Ga-I-5, Hg-I-45, In-I-37, In-I-45, Mn-I-5, Mn-I-25, Mn-I-45, Mn-I-51, Mn-I-80, Ni-I-17, Ni-I-18, Ni-I-19, Ni-I-20, Ni-I-21, Ni-I-22, Ni-I-25, Ni-I-26, Ni-I-28, Ni-I-30, Ni-I-31, Ni-I-33, Ni-I-34, Ni-I-39, Ni-I-41, Ni-I-43, Ni-I-50, Ni-I-51, Ni-I-55, Ni-I-56, Ni-I-57, Ni-I-59, Ni-I-61, Ni-I-64, Ni-I-65, Ni-I-68, Ni-I-69, Ni-I-71, Ni-I-72, Ni-I-73, Ni-I-75, Ni-I-76, Ni-I-79, Ni-I-80, Ni-I-82, Ni-I-91, Ni-I-96, Ni-I-98, Ni-I-102, Ni-I-103, Pb-I-94, Pb-I-96, Pb-I-97, Pt-I-94, Rh-I-36, Sb-I-5, Sn-I-5, Tl-I-5, Tl-I-45, V-I-28, V-I-45, W-I-45, Zn-I-5, Zn-I-21, Zn-I-25, Zn-I-35, Zn-I-42, Zn-I-45, Zn-I-48, Zn-I-50, Zn-I-51, Zn-I-58, Zn-I-68, Zn-I-72, Zn-I-74, Zn-I-96, Zn-I-97, Zn-I-98, Zn-I-101

Magnetic Separation: Ag-M-6, Ag-M-20, Ag-M-22, Al-M-3, Al-M-13, Al-M-14, Al-M-16, Al-M-21, As-M-19, Au-M-6, Bi-M-19, Bronze-M-24, Cd-M-9,

Cr-P-23,   Cr-P-24,   Cr-P-25,   Cr-P-27,   Cr-P-32,
Cr-P-54,   Cr-P-66,   Cr-P-67,   Cr-P-68,   Cr-P-84,
Cr-P-110, Cu-P-15,   Cu-P-17,   Cu-P-19,   Cu-P-23,
Cu-P-25,   Cu-P-28,   Cu-P-30,   Cu-P-33,   Cu-P-34,
Cu-P-36,   Cu-P-37,   Cu-P-38,   Cu-P-39,   Cu-P-40,
Cu-P-43,   Cu-P-44,   Cu-P-45,   Cu-P-46,   Cu-P-47,
Cu-P-49,   Cu-P-51,   Cu-P-52,   Cu-P-53,   Cu-P-54,
Cu-P-56,   Cu-P-57,   Cu-P-59,   Cu-P-62,   Cu-P-66,
Cu-P-68,   Cu-P-70,   Cu-P-71,   Cu-P-72,   Cu-P-73,
Cu-P-74,   Cu-P-76,   Cu-P-79,   Cu-P-81,   Cu-P-85,
Cu-P-86,   Cu-P-87,   Cu-P-88,   Cu-P-91,   Cu-P-93,
Cu-P-95, Cu-P-105, Cu-P-106, Cu-P-109, Cu-P-113,
Fe-P-15,   Fe-P-24,   Fe-P-36,   Fe-P-37,   Fe-P-46,
Fe-P-58,   Fe-P-64,   Fe-P-84, Fe-P-110, Fe-P-111,
Hg-P-31,   Hg-P-54,   Mg-P-37,   Mn-P-37,   Mn-P-53,
Mn-P-71,   Mn-P-94,   Mo-P-52,   Mo-P-53,   Mo-P-67,
Ni-P-15,   Ni-P-16,   Ni-P-17,   Ni-P-18,   Ni-P-21,
Ni-P-23,   Ni-P-24,   Ni-P-25,   Ni-P-26,   Ni-P-27,
Ni-P-31,   Ni-P-35,   Ni-P-36,   Ni-P-41,   Ni-P-48,
Ni-P-50,   Ni-P-51,   Ni-P-58,   Ni-P-59,   Ni-P-61,
Ni-P-63,   Ni-P-65,   Ni-P-66,   Ni-P-67,   Ni-P-68,
Ni-P-72,   Ni-P-73,   Ni-P-77,   Ni-P-84,   Ni-P-86,
Ni-P-91,   Ni-P-94, Ni-P-107, Ni-P-108, Ni-P-114,
Pb-P-19,   Pb-P-20,   Pb-P-28,   Pb-P-31,   Pb-P-33,
Pb-P-54,   Pb-P-71,   Pb-P-74,   Pb-P-76,   Pb-P-80,
Pb-P-87, Pb-P-106, Pb-P-111, Pb-P-112,   Pd-P-31,
Pd-P-39,   Pt-P-54,   Rh-P-31,   Rh-P-54,   Sb-P-54,
Sn-P-54,   Sn-P-65,   Sn-P-75,   Sn-P-87,   Ti-P-84,
V-P-58, V-P-63, W-P-59, W-P-67, Zn-P-19, Zn-P-20,
Zn-P-28,   Zn-P-36,   Zn-P-37,   Zn-P-45,   Zn-P-47,
Zn-P-56,   Zn-P-60,   Zn-P-64,   Zn-P-65,   Zn-P-66,
Zn-P-68,   Zn-P-71,   Zn-P-72,   Zn-P-73,   Zn-P-80,
Zn-P-112

Pyrometallurgical: Ag-H-6, Ag-H-7, Ag-H-21, Ag-H-26,
Ag-H-28, Al-H-14, Al-H-23, Al-H-36, As-H-19,
Au-H-7, Au-H-12, Au-H-31, Cd-H-15, Co-H-5,
Co-H-8, Co-H-10, Co-H-14, Co-H-30, Co-H-38,

Mo-S-33, Mo-S-34, Mo-S-35, Mo-S-47, Mo-S-56,
Mo-S-59, Mo-S-66, Ni-S-10, Ni-S-12, Ni-S-15,
Ni-S-18, Ni-S-19, Ni-S-23, Ni-S-24, Ni-S-33,
Ni-S-34, Ni-S-35, Ni-S-42, Ni-S-43, Ni-S-44,
Ni-S-45, Ni-S-46, Ni-S-47, Ni-S-48, Ni-S-49,
Ni-S-50, Ni-S-51, Ni-S-52, Ni-S-54, Ni-S-55,
Ni-S-56, Ni-S-57, Ni-S-58, Ni-S-59, Ni-S-60,
Ni-S-61, Ni-S-62, Ni-S-63, Ni-S-64, Ni-S-65,
Ni-S-66, Ni-S-67, Ni-S-68, Ni-S-69, Ni-S-70,
Ni-S-71, Ni-S-72, Pb-S-49, Pb-S-51, Pd-S-64,
Pd-S-67, Sn-S-52, U-S-38, U-S-64, U-S-67, V-S-26,
V-S-33, V-S-34, V-S-35, V-S-52, V-S-53, V-S-55,
V-S-56, W-S-33, W-S-47, W-S-56, W-S-59, W-S-66,
Zn-S-37, Zn-S-38, Zn-S-40, Zn-S-41, Zn-S-42,
Zn-S-48, Zn-S-49, Zn-S-50, Zn-S-51, Zn-S-52,
Zn-S-54, Zn-S-58, Zn-S-61, Zn-S-66, Zn-S-69

Solubilization: Ag-SO-31,   Ag-SO-33,   Ag-SO-37,
Al-SO-33,   As-SO-36,   As-SO-37,   Au-SO-14,
Au-SO-15,   Au-SO-16,   Au-SO-31,   Au-SO-33,
Au-SO-35,   Au-SO-47,   Ba-SO-14,   Cd-SO-14,
Cd-SO-36,   Cd-SO-37,   Cd-SO-40,   Ce-SO-14,
Co-SO-3, Co-SO-5, Co-SO-6, Co-SO-14, Co-SO-24,
Co-SO-38,   Co-SO-42,   Co-SO-44,   Cr-SO-14,
Cr-SO-15, Cr-SO-34, Cr-SO-47, Cu-SO-5, Cu-SO-7,
Cu-SO-12,   Cu-SO-14,   Cu-SO-15,   Cu-SO-16,
Cu-SO-24,   Cu-SO-25,   Cu-SO-27,   Cu-SO-28,
Cu-SO-29,   Cu-SO-30,   Cu-SO-31,   Cu-SO-32,
Cu-SO-33,   Cu-SO-34,   Cu-SO-36,   Cu-SO-37,
Cu-SO-39,   Cu-SO-41,   Cu-SO-45,   Cu-SO-48,
Fe-SO-15, Fe-SO-24, Fe-SO-25, Fe-SO-33, Fe-SO-37,
Li-SO-23,   Mg-SO-14,   Mg-SO-15,   Mn-SO-14,
Mn-SO-15,   Mo-SO-14,   Mo-SO-44,   Nb-SO-14,
Ni-SO-5, Ni-SO-6, Ni-SO-7, Ni-SO-11, Ni-SO-14,
Ni-SO-15,   Ni-SO-24,   Ni-SO-25,   Ni-SO-30,
Ni-SO-34,   Ni-SO-38,   Ni-SO-40,   Ni-SO-42,
Ni-SO-43,   Ni-SO-44,   Ni-SO-45,   Ni-SO-46,
Ni-SO-47,   Pb-SO-15,   Pb-SO-16,   Pb-SO-27,

Index by Metal

Key: aa = Metal; bb-nnn: bb = Process; nnn = Reference

Process:
    A—Adsorption;   B—Biological;   C—Cementation;   E—
    Electrowinning;   F—Flotation;   I—Ion   Exchange;   M—
    Magnetic Separation; MS—Membrane Separation; PS—
    Precipitation Separation; H—Pyrometallurgical; S—Sol-
    vent   Extraction;   SO—Solubilization;   SC—Spent
    Catalyst

H–22, H–23, H–24, H–29, H–30, H–31, H–33, H–35,
H–37, H–38, H–39, H–40, H–42, H–43, I–19, I–21, I–22,
I–24, I–25, I–27, I–29, I–32, I–35, I–38, I–40, I–41, I–42,
I–43, I–44, I–45, I–46, I–47, I–48, I–49, I–51, I–53, I–54,
I–59, I–60, I–61, I–62, I–63, I–66, I–67, I–68, I–69, I–71,
I–72, I–74, I–76, I–78, I–81, I–82, I–86, I–88, I–89, I–91,
I–92, I–97, I–98, I–99, I–100, I–101, M–3, M–4, M–5,
M–7, M–8, M–9, M–10, M–12, M–13, M–14, M–15,
M–16, M–17, M–18, M–19, M–21, M–22, M–24, MS–14,
MS–15, MS–17, MS–19, MS–20, MS–22, MS–30, MS–31,
MS–34, MS–36, MS–37, MS–40, MS–41, MS–44, MS–45,
MS–46, MS–48, MS–49, MS–50, P–15, P–17, P–19, P–23,
P–25, P–28, P–30, P–33, P–34, P–36, P–37, P–38, P–39,
P–40, P–43, P–44, P–45, P–46, P–47, P–49, P–51, P–52,
P–53, P–54, P–56, P–57, P–59, P–62, P–66, P–68, P–70,
P–71, P–72, P–73, P–74, P–76, P–79, P–81, P–85, P–86,
P–87, P–88, P–91, P–93, P–95, P–105, P–106, P–109,
P–113, S–10, S–17, S–19, S–24, S–35, S–36, S–37, S–38,
S–40, S–41, S–42, S–44, S–45, S–47, S–48, S–49, S–50,
S–51, S–53, S–54, S–55, S–58, S–61, S–64, S–66, S–67,
S–69, S–70, SC–2, SC–19, SO–5, SO–7, SO–12, SO–14,
SO–15, SO–16, SO–24, SO–25, SO–27, SO–28, SO–29,
SO–30, SO–31, SO–32, SO–33, SO–34, SO–36, SO–37,
SO–39, SO–41, SO–45, SO–48

Fe: A–12, B–2, C–1, C–2, C–3, C–6, C–7, C–8, C–9, C–10,
C–11, C–17, C–18, C–19, C–22, F–9, F–12, F–13, F–41,
F–44, F–48, F–53, F–54, F–59, H–5, H–20, H–21, H–25,
H–26, H–30, H–38, H–42, I–5, I–18, I–51, I–56, I–68,
I–80, I–96, I–97, M–1, M–2, M–4, M–9, M–21, M–22,
MS–41, MS–51, P–15, P–24, P–36, P–37, P–46, P–58,
P–64, P–84, P–110, P–111, S–7, S–8, S–40, S–41, S–47,
S–52, S–64, SC–15, SO–15, SO–24, SO–25, SO–33,
SO–37

Ga: C–1, I–5

P–26, P–27, P–31, P–35, P–36, P–41, P–48, P–50, P–51, P–58, P–59, P–61, P–63, P–65, P–66, P–67, P–68, P–72, P–73, P–77, P–84, P–86, P–91, P–94, P–107, P–108, P–114, S–10, S–12, S–15, S–18, S–19, S–23, S–24, S–33, S–34, S–35, S–42, S–43, S–44, S–45, S–46, S–47, S–48, S–49, S–50, S–51, S–52, S–54, S–55, S–56, S–57, S–58, S–59, S–60, S–61, S–62, S–63, S–64, S–65, S–66, S–67, S–68, S–69, S–70, S–71, S–72, SC–3, SC–4, SC–5, SC–6, SC–7, SC–9, SC–10, SC–11, SC–12, SC–13, SC–16, SC–19, SC–22, SO–5, SO–6, SO–7, SO–11, SO–14, SO–15, SO–24, SO–25, SO–30, SO–34, SO–38, SO–40, SO–42, SO–43, SO–44, SO–45, SO–46, SO–47

Pb: A–17, B–2, B–5, B–14, B–15, B–16, C–1, C–15, E–39, E–54, E–72, E–78, E–114, E–123, E–124, E–135, F–18, F–24, F–33, F–52, F–54, F–55, F–60, F–75, H–23, H–38, I–94, I–96, I–97, M–3, M–7, M–8, M–13, M–14, M–17, M–19, MS–13, MS–46, P–19, P–20, P–28, P–31, P–33, P–54, P–71, P–74, P–76, P–80, P–87, P–106, P–111, P–112, S–49, S–51, SO–15, SO–16, SO–27, SO–37, SO–47

Pd: A–8, A–14, A–19, C–14, C–16, E–86, E–97, P–31, P–39, S–64, S–67, SC–8, SC–14, SC–17, SC–21, SO–18

Pt: I–94, M–15, P–54, SC–8, SC–14, SC–17, SC–21, SO–18

^{242}Pu: B–12

Ra: B–2, B–3, SO–41

Rh: I–36, M–15, P–31, P–54, SC–8, SC–16, SC–25, SO–18

Sb: A–17, C–12, H–19, I–5, P–54

Se: B–2, B–3, H–21

Sn: A–17, B–2, C–12, E–38, E–43, E–101, E–106, E–112, E–114, F–8, F–60, H–23, H–38, I–5, M–8, M–24, P–54, P–65, P–75, P–87, S–52, SO–47, SO–48